Diseases Ornamental in Plant

Diseases Ornamental in Plant

Dr. N.K. Rajan

AMIGA PRESS. INC
Delhi -110051

AMIGA PRESS. INC
H.O :- 28-A, Indra Park, Chander Nagar
Delhi :- 110051 (India)
Tel :- 011-22504028, +91-98736-57000
Email :- amigapress@gmail.com
amigaeditorial@gmail.com
amigapub@gmail.com
Website :- www.amigapress@co.in

This edition is published by Amiga Press.Inc

ISBN : 978-93-84533-33-5

First Published 2018

PRINTED : REPLIKA PRESS PVT. LTD.

Preface

Plant diseases are usually caused by fungi, bacteria and viruses. Also there are other diseases which are caused by adverse environmental conditions. Plant diseases are known from times preceding the earliest writings. Fossil evidence indicates that plants were affected by disease 250 million years ago.

Bacterial diseases are generated by bacteria in the plants. These are very small organisms, not seen by the naked eyes. The bacteria are disseminated in infected seeds and travel from plant to plant and field through blowing winds, splashing rains, sprinkler irrigation, surface-drainage water, insects, birds, animals and human beings etc. The diseases typically develop, when contaminated seeds are planted in fields with a history of the diseases and when the climate is consistently hot and wet or humid. Disease plants may be dwarfed, with the upper leaves crinkled and mottled. Dark green, water-soaked areas may also be visible on the stems. Young seedlings, infected with halo blight develop a chlorosis, which may be general or confined to the young trifoliate leaves at the top of the plant.

The most common diseases encountered in the landscape are foliar leaf spots and blights. The key to managing these diseases lies in the environmental conditions that favor disease development. Foliar pathogens are typically active with high moisture availability and lengthened periods of leaf wetness. Factors that extend leaf wetness periods are the driving forces of foliar disease.

By irrigating early in the morning, the water will wash off leaf-bound dew that typically is laden with nutrient-rich leaf exudates (a food source for foliar fungal pathogens). You also benefit from the low wind speeds at that time and lack of customers or patrons on the site. On large sites where you cannot irrigate the entire landscape in the early morning hours, consider locating plants that are susceptible to disease in locations that favor rapid drying of the foliage (e.g. eastern side facing sun).

Factors which affect Plant diseases are microorganisms, including fungi, bacteria, viruses, mycoplasmas, etc. or may be incited by physiological causes including high or low temperatures, lack or excess of soil moisture and aeration, deficiency or excess of plant nutrients, soil acidity or alkalinity, etc. Factors that limit the rate of disease development are the relatively low amounts of inoculum in the leg stage and the paucity of healthy plants available to the inoculum in the stationary stage.

The causative agents of disease in green plants number in a tens of thousands and include almost every form of life. But primary agents of disease may also be inanimate. Thus nonliving (abiotic) agents of disease include mineral deficiencies and excesses, air pollutants, biologically produced toxicants, improperly used pesticidal chemicals, and such other environmental factors as wind, water, temperature, and sunlight. Nonliving things certainly qualify as primary agents of disease; they continuously irritate plant cells and tissues; they are harmful to the physiological processes of the plant; and they evoke pathological responses that manifest as the symptoms characteristic of the several diseases. But the abiotic agents of disease in plants. The abiotic agents of plant disease are termed noninfectious, and the diseases they cause are termed noninfectious diseases.

The book fulfil not only the need of the students to find literature on the diseases and other pathological conditions difficult to obtain and access, but also provide complete systematic treatment of the subject from their point of view.

—Editor

Contents

1

Introduction

When an insect pest affects any crop, by critical examination of the affected plant specimen with an ordinary hand lense, several minute insects like aphids, jassid hoppers, flies or small larvae, etc. can be seen moving and feeding on the plant parts. In case of a plant disease, however, no such minute insects are seen on the affected plant parts but only the symptoms like leaf spots, blightening of leaves or growing shoots, writing of entire plant, root rot or fruit rot, curling and dwarfing, etc. are seen to have developed. Insect pests cause losses to plants by two ways. Some insects eat away the plant parts while others only suck the plant sap. The former are called 'biting' or 'chewing' insects while the latter are called as 'sucking' insects. In case of diseases on the other hand, the microorganisms responsible for causing the most diseases utilize the plant food available in different parts viz. Leaves, flowers, fruits, stems, roots, etc. and grow inside or on the surface of these plant parts under favourable conditions and develop different kinds of symptoms on the affected parts. Neverthless, in certain cases besides such involvement of any kind of microorganism, other unfavourable factors relating to soil, climate or nutrition, etc. and sometimes certain insects also develop symptoms resembling to those caused by different microorganisms.

DESIGNING A GARDEN

Speaking broadly, a person will get from a garden what he puts into it; and it is of the first importance, therefore, that a clear conception of the work be formulated at the outset. I do not mean to say that the garden will always turn out what it was desired that it should be; but the failure to turn out properly is usually some fault in the first plan or some neglect in execution.

Sometimes the disappointment in an ornamental garden is a result of confusion of ideas as to what a garden is for. The garden ought to express the feeling of the different months. The failing leaves and expended plants are therefore to be looked on, to some extent at least, as the natural order and destiny of a good garden.

These attributes are well exhibited in the vegetable-garden. In the spring, the vegetable-garden is a model of neatness and precision. The rows are straight. There are no missing plants. The earth is mellow and fresh. Weeds are absent. One takes his friends to the garden, and he makes pictures of it. By late June or early July, the plants have begun to sprawl and to get out of shape. The bugs have taken some of them. The rows are no longer trim and precise. The earth is hot and dry. The weeds are making headway. By August and September, the garden has lost its early regularity and freshness. The camera is put aside. The visitors are not taken to it: the gardener prefers to go alone to find the melon or the tomatoes, and he comes away as soon as he has secured his product. Now, as a matter of fact, the garden has been going through its regular seasonal growth. It is natural that it become ragged.

There are two types of outdoor gardening in which the progress of the season is not definitely expressed,—in the carpet-bedding kind, and in the subtropical kind. I hope that my reader will get a clear distinction in these matters, for it is exceedingly important. The carpet-bedding gardening is the making of figure-beds in house-leeks and achyranthes and coleus and sanitalia, and other things that can be grown in compact masses and possibly sheared to keep them within place and bounds; the reader sees these beds in perfection in some of the parks and about florists' establishments; he will understand at once that they are not meant in any way to express the season, for the difference between them in September and June is only that they may be more perfect in September.

The subtropical gardening is the planting out of house-grown stuff, in order to produce given effects, of such plants as palms, dracenas, crotons, caladiums, papyrus, together with such luxuriant things as dahlias and cannas and large ornamental grasses and castor beans; these plants are to produce effects quite foreign to the expression of a northern landscape, and they are usually at their best and are most luxuriant when overtaken by the fall frosts.

Now, the home gardener usually relies on plants that more or less come and go with the seasons. He pieces out and extends the season, to be sure; but a garden with pansies, pinks, sweet william, roses, sweet peas, petunias, marigolds, salpiglossis, sweet sultan, poppies, zinnias, asters, cosmos, and the rest, is a progress-of-the-season garden, nevertheless; and if it is a garden of herbaceous perennials, it still more completely expresses the time-of-year.

My reader will now consider, perhaps, whether he would have his garden accent and heighten his natural year from spring to fall, or whether he desires to thrust into his year a feeling of another order of vegetation. Either is allowable; but the gardener should distinguish at the outset.

There are two ways by which the garden year may be extended: by planting things that bloom very late in fall and others that bloom very early in spring; by using freely, in the backgrounds, of bushes and trees that have interesting winter characters.

One cannot expect satisfaction in the planting and developing of a home area unless he has a clear conception of what is to be done. The homemaker should develop his plan before he attempts to develop his place. He must study the various subdivisions in order that the premises may meet all his needs. He should determine the locations of the leading features of the place and the relative importance to be given to the various parts of it,—as of the landscape parts, the ornamental areas, the vegetable-garden, and the fruit plantation.

It is desirable to have a definite plan on paper (drawn to scale) for the location of the leading features of the place. These features are the residence, the out-houses, the walks and drives, the service areas (as clothes yards), the border planting, flower-garden, vegetable-garden, and fruit-garden. It should not be expected that the map plan can be followed in every detail, but it will serve as a general guide; and if it is made on a large enough scale, the different kinds of plants can be located in their proper positions, and a record of the place be kept.

These remarks are illustrated in the aesthetics of landscape gardening. It is the artist's one desire to make pictures in the landscape. This is done in two ways: by the form of plantations, and by the use of vistas. He will throw his plantations into such positions that open and yet more or less confined areas of greensward are presented to the observer at various points. This picture-like opening is nearly or quite devoid of small or individual objects, which usually destroy the unity of such areas and are meaningless in themselves. A vista is a narrow opening or view between plantations to a distant landscape. It cuts up the broad horizon into portions that are readily cognizable. It frames parts of the country-side. The verdurous sides of the planting are the sides of the frame; the foreground is the bottom, and the sky is the top. It is of the utmost importance that good views be left or secured from the best windows of the house (not forgetting the kitchen window); in fact, the placing of the house may often be determined by the views that may be appropriated.

If a landscape is a picture, it must have a canvas. This canvas is the greensward. Upon this, the artist paints with tree and bush and flower as the painter does upon his canvas with brush and pigments. The opportunity for artistic composition and design is nowhere so great as in the landscape garden, because no other art has such a limitless field for the expression of its emotions. It is not strange, if this be true, that there have been few great landscape gardeners, and that, falling short of art, the landscape gardener too often works in the sphere of the artisan. There can be no rules for landscape gardening, any more than there can be for painting or sculpture. The operator may be taught how to hold the brush or strike the chisel or plant the tree, but he remains an operator; the art is intellectual and emotional and will not confine itself in precepts.

The making of a good and spacious lawn, then, is the very first practical consideration in a landscape garden.

The lawn provided, the gardener conceives what is the dominant and central feature in the place, and then throws the entire premises into subordination to this feature. In home grounds this central feature is the house. To scatter trees and bushes over the area defeats the fundamental purpose of the place,—the purpose to make every part of the grounds lead up to the home and to accentuate its homelikeness.

Of course these remarks about the lawn are meant for those countries where greensward is the natural ground cover. In the South and in arid countries, greensward is not the prevailing feature of the landscape, and in these regions the landscape design may take on a wholly different character, if the work is to be nature-like. We have not yet developed other conceptions of landscape work to any perfect extent, and we inject the English greensward treatment even into deserts. We may look for the time when a brown landscape garden may be made in a brown country, and it may be good art not to attempt a broad open centre in regions in which undergrowth rather than sod is the natural ground cover. In parts of the United States we are developing a good Spanish-American architecture, perhaps we may develop a recognized comparable landscape treatment as an artistic expression.

GARDENING/PLANTS

There is a knack in the successful handling of plants that it is impossible to describe in print. All persons can improve their practice through diligent reading of useful gardening literature, but no amount of reading and advice will make a good gardener of a person who does not love to dig in a garden or who does not have a care for plants just because they are plants. To grow a plant well, one must learn its natural habits. Some persons learn this as if by intuition, acquiring the knowledge from close discrimination of the behaviour of the plant. Often they are themselves unconscious of this knack of knowing what will make the plant to thrive; but it is not at all necessary to have such an intuitive judgment to enable one to be even more than a fairly good gardener. Diligent attention to the plant's habits and requirements, and a real regard for the plant's welfare, will make any person a successful plant-grower.

Important Facts

To learn before you grow;

- Whether the plant matures in the first, second, third, or subsequent years; and when it naturally begins to fail.
- The time of the year or season in which it normally grows, blooms, or fruits; and whether it can be forced at other seasons.
- Whether it prefers a situation dry or moist or wet, hot or cool, sunny or shady.
- Its preferences as to soil, whether very rich or only moderately rich, sand or loam, or peat or clay.
- Its hardiness as to frost, wind, drought, heat.
- Whether it has any special requirements as to germination, and whether it transplants well.
- Whether it is specially liable to attack by insects or disease.
- Whether it has a special inability to grow two years in succession on the same land.

To Prepare to Grow

- Guarding from all insects and diseases; and also from cats and chickens and dogs; and likewise from rabbits and mice.
- Protecting from weeds.
- Pruning, in the case of fruit trees and bushes, and also of ornamental woody plants on occasion, and sometimes even of annual herbs.
- Staking and tying, particularly of sprawly garden flowers.
- Persistent picking of seed pods or dead flowers from flower plants, in order to conserve the strength of the plant and to prolong its season of bloom.
- Watering in dry weather (but not sprinkling or dribbling).
- Thorough winter protecting of plants that need it.
- Removing dead leaves, broken branches, weak and sickly plants, and otherwise keeping the place tidy and trim.

Sowing Seeds

Prepare the surface earth well, to make a good seed-bed. Plant when the ground is moist, if possible, and preferably just before a rain if the soil is of such character that it will not bake. For shallow-planted seeds, firm the earth above them by walking over the row or by patting it down with a hoe. Special care should be exercised not to sow very small and slow-germinating seeds, as celery, carrot, onion, in poorly prepared soil or in ground that bakes. With such seeds it is well to sow seeds of radish or turnip, for these germinate quickly and break the crust, and also mark the row so that tillage may be begun before the regular-crop seeds are up.

Land may be prevented from baking over the seeds by scattering a very thin layer of fine litter, as chaff, or of sifted moss or mold, over the row. A board is sometimes laid on the row to retain the moisture, but it must be lifted gradually just as soon as the plants begin to break the ground, or the plants will be greatly injured. Whenever practicable, seed-beds of celery and other slow-germinating seeds should be shaded. If the beds are watered, be careful that the soil is not packed by the force of the water or baked by the sun. In thickly sown seed-beds, thin or transplant the plants as soon as they have made their first true leaves.

For most home-grounds, seeds may be sown by hand, but for large areas of one crop, one of the many kinds of seed-sowers may be used. The particular methods of sowing seeds are usually specified in the seed catalogues, if other than ordinary treatment is required. The sled-markers open a furrow of sufficient depth for the planting of most seeds. If marker furrows are not available, a furrow may be opened with a hoe for such deep-planted seeds as peas and sweet peas, or by a trowel or end of a rakestale for smaller seeds. In narrow beds or boxes, a stick or ruler may be used for opening creases to receive the seeds.

The depth at which seeds are to be planted varies with the kind, the soil and its preparation, the season, and whether they are planted in the open or in the house. In boxes and under glass, it is a good rule that the seed be sown at a depth equal to twice its own diameter, but deeper sowing is usually necessary out of doors, particularly in hot and dry weather. Strong and hardy seeds, as peas, sweet peas, large fruit-tree seeds, may be planted three to six inches deep. Tender seeds, that are injured by cold and wet, may be planted after the ground is settled and warm at a greater depth than before that season. As a rule, nothing is gained by sowing tender seeds before the weather is thoroughly settled and the ground warm.

Propagating by Cuttings

Many common plants are propagated by cuttings rather than by seeds, particularly when it is desired to increase a particular variety.

Cuttings are parts of plants inserted in soil or water with the intention that they shall grow and make new plants. They are of various kinds. They may be classified, with reference to the age of the wood or tissue, into two classes; viz. those made from perfectly hard or dormant wood (taken from the winter twigs of trees and bushes), and those made from more or less immature or growing wood. They may be classified again in respect to the part of the plants from which they are taken, as root-cuttings, tuber-cuttings (as the ordinary "seed" planted for potatoes), stem-cuttings, and leaf-cuttings.

Dormant Stem-cuttings

Dormant-wood cuttings are used for grapes, currants, gooseberries, willows, poplars, and many other kinds of soft-wooded trees and shrubs. Such cuttings are ordinarily taken in fall or winter, but cut into the proper lengths and then buried in sand or moss where they do not freeze, in order that the lower end may heal over or callous. In the spring these cuttings are set in the ground, preferably in a rather sandy and well-drained place.

Usually, hardwood cuttings are made with two to four joints or buds, and when they are planted, only the upper bud projects above the ground. They may be planted erect or somewhat slanting. In order that the cutting may reach down to moist earth, it is desirable that it should not be less than 6 in. long; and it is sometimes better if it is 8 to 12 in. If the wood is short-jointed, there may be several buds on a cutting of this length; and in order to prevent too many shoots from arising from these buds the lowermost buds are often cut out. Roots will start as readily if the lower buds are removed, since the buds grow into shoots and not into roots.

Cuttings of currants, grapes, gooseberries, and the like may be set in rows that are far enough apart to admit of easy tillage either with horse or hand tools, and the cuttings may be placed 3 to 8 in. apart in the row. The English varieties of gooseberries, considerably grown in this country, do not propagate readily from cuttings.

After the cuttings have grown one season, the plants are usually transplanted and given more room for the second year's growth, after which time they are ready to be set in permanent plantations. In some cases, the plants are set at the end of the first year; but two-year plants are stronger and usually preferable.

Root Cuttings

Root-cuttings are used for blackberries, raspberries, and a few other things. They are ordinarily made of roots from the size of a lead pencil to one's little finger, and are cut in lengths from 3 to 5 in. long. The cuttings are stored the same as stem-cuttings and allowed to callous. In the spring they are planted in a horizontal or nearly horizontal position in moist sandy soil, being entirely covered to a depth of 1 or 2 in.

Green Cuttings

Softwood or greenwood cuttings are usually made of wood that is mature enough to break when it is bent sharply. When the wood is so soft that it will bend and not break, it is too immature, in the majority of plants, for the making of good cuttings.

One to two joints is the proper length of a greenwood cutting. If of two joints, the lower leaves should be cut off and the upper leaves cut in two so that they do not present their entire surface to the air and thereby evaporate the plant juices too rapidly. If the cutting is of only one joint, the lower end is usually cut just above a joint. In either case, the cuttings are usually inserted in sand or well-washed gravel, nearly or quite up to the leaves. Keep the bed uniformly moist throughout its depth, but avoid any soil which holds so much moisture that it becomes muddy and sour. These cuttings should be shaded until they begin to emit their roots. Coleus, geraniums, fuchsias, carnations, and nearly all the common greenhouse and house plants, are propagated by these cuttings or slips.

Leaf Cuttings

Leaf-cuttings are often used for the fancy-leaved begonias, gloxinias, and a few other plants. The young plant usually arises most readily from the leaf-stalk or petiole. The leaf, therefore, is inserted into the ground much as a green cutting is. Begonia leaves will throw out young plants from the main ribs when these veins or ribs are cut. Therefore, well-grown and firm begonia leaves are sometimes laid flat on the sand and the main veins cut; then the leaf is weighted down with pebbles or pegs so that these cut surfaces come into intimate contact with the soil beneath. The usual way, however, is to cut a triangular piece of the leaf and insert the tip in sand. So long as the cutting is alive, do not be discouraged, even if it does not start.

General Treatment of Cuttings

In the growing of all greenwood and leaf-cuttings, it is well to remember that they should have a gentle bottom heat; the soil should be such that it will hold moisture and yet not remain wet; the air about the tops should not become close and stagnant, else the plants will damp off; and the tops should be shaded for a time. In order to control all the conditions, such cuttings are grown under cover, as in a greenhouse, coldframe, or a box in the residence window.

An excellent method of starting cuttings in the living room is to make a double pot. Inside a 6-in. pot set a 4-in. pot. Fill the bottom, _a,_ with gravel or bits of brick, for drainage. Plug the hole in the inside pot. Fill the spaces between, _c,_ with earth, and in this set the cuttings. Water may be poured into the inner pot, _b,_ to supply the moisture.

CAUSES OF PLANT DISEASES

Two types of factors cause plant diseases viz.-

Parasitic Causes

These include different types of microorganisms. Infection and development of microorganisms inside or on the plant surface, however, necessarily require suitable climate and a susceptible host plant variety. In majority of the plant diseases following types of microorganisms are encountered as the causal agents-

Fungi:: In the rainy season, when high humidity prevails for longer duration's leather goods, fruits and vegetables, bread or other organic matter left in unventilated places usually develop white, gray, yellow or bluish green powdery growth on their surface. This growth consists of different kinds of living microorganisms called Fungi (singular-fungus) and thousands of such fungal species exists in nature. Many of them survive and grow on dead decaying organic matter of plant or animal origin; while some others grow only on living tissues of plants or animals. The latter, therefore, are called 'parasites' while the former are termed as 'saprophytes.' The parasitic fungi are thus responsible for causing different diseases to plants and animals. Fungal body consists of minute threadlike structures called 'mycelium' which may be unicellular or it may be multicellular Under favourable conditions, they grow very fast and in a few days produce innumerable microscopic seedlike reproductive structures called 'spores.' Wind, rain or irrigation water and insects, etc easily carry these spores from infected plants to healthy plants and thus are responsible for rapid spread of the disease. Under unfavourable conditions of climate (temperature and humidity) and food the fungi develop different types of resting structures which remain in dormant condition and start their activity afresh as soon as favourable conditions prevail. The lifecycle of fungi, thus, consists mainly of two phases viz.:-

1. Resting stage-In many fungi, this stage is in the form of different types of spores developed by sexual methods or it may be in the form of thick-walled mycelium or its thick-walled individual cells. These resting or dormant structures are highly resistant to adverse conditions like very high or low temperatures, lack of food and water, etc. for varying periods ranging from few days to some years.
2. Active Stage:-As soon as the conditions of climate and food become favourable the resting structures restart their ability by producing fresh mycelium which causes the 'primary infection' of plant diseases, the fungal mycelium growing fast inside or on the host plant tissues. Further spread of the causal fungus and the 'secondary spread' of the disease occurs mainly by innumerable spores rapidly developed by asexual methods and carried by wind, water, insects, implements, etc. from diseased plants to healthy ones. Most of the severe plant diseases of different crops are caused mainly by different kinds or 'species' of fungi and are responsible for great economic losses. The common symptoms of such diseases can be short-listed as leafspots, mildews, blights, rots and wilts, etc.

Bacteria: Bacteria (singular-bacterium) are microscopic unicellular organisms with their shapes varying from round to oval or short rods They do not produce reproductive structures like fungi but the bacterial cells themselves divide from one to two, from two to four and so on. The dispersion commonly occurs by water, insects, seeds and implements, etc. Bacteria usually are not carried by wind. Plant parasitic bacteria have no specialized resting stage but under unfavourable conditions some of the several bacterial cells remain dormant inside the affected parts or even in the dead and dried plant tissues for some days or months and start their activity when conditions become favourable. As compared to fungal plant diseases severe plant diseases caused by bacteria are less.

The common symptoms developed by bacterial plant diseases are leafspots, blights, fruits and vegetables, rots, cankers and wilts.

Viruses: Viruses are not mentioned as true microorganisms because of their characters in between living and non-living entities and their sub-microscopic size. They are therefore, treated as ultra-microscopic 'particles'. They also do not ingest any food and do not grow in size and number like true microorganisms. They survive only inside the living individual cells of plants or animal tissues. When once they come in contact with the cell sap or contents of the cells, several such 'particles' are naturally developed from the cell sap itself only due to the presence or contact of the original particle(s). Due to this character, the viruses are treated as ' contagious'. Although, their body is composed of organic matter it lacks in certain criteria to donate the characters of a true living organisms. Viruses are mainly carried from diseased plants to healthy ones by different kinds of insects such as aphids, jassid hoppers, flies and mites, etc.

Besides these agents some viruses are also carried by nematodes or by direct contact through wounds and injuries. When a viral disease infects once a plant, it never recovers for the viruses are practically immune to any chemical treatment. The symptoms of viral diseases are characteristic and they include chlorosis, mosaic, curling, dwarfing and sometimes leaf spots or specks. The affected plants usually do not die immediately but remain nearly unproductive.

Nematodes: Nematodes are microscopic wormlike organisms reproducing sexually by production of eggs. The males and females differ in their shape. These organisms live either freely in soil or inside plant roots. Plant parasitic nematodes damage the plants by feeding on the roots either externally or internally.

The nematodes living freely in the soil may be stationary or may travel short distances. Water, soil and implements, etc. are the factors responsible for spread of nematodes from one place to another. Some nematodes enter the plant roots and develop knots of different sizes on the roots while some other kinds of nematodes remain outside the roots and feed on the roots externally. The root injuries caused by such external feeders serve as avenues for entry of parasitic fungi and bacteria, which cause further damage to the plants. The plants infected by nematodes loose vigour and turgidity, appear sick or unhealthy, lack in productivity and may die eventually.

Non-Parasitic causes: The disease symptoms resembling to those produced by different microorganisms or viruses are sometimes developed by other factors also. Such factors include-

1. Unfavourable soil conditions-illdrained soil with high salt contents.
2. Imbalanced nutrients-high or low proportion of macro and/or micronutrients.
3. Chemicals or Hormones:-Some chemicals used for control of pests or diseases may prove to be phytotoxic to plants. Chemicals or hormones used in excess proportions may also prove harmful.
4. Water-salty, with high or low pH, polluted, inadequate or excess water develops undesired symptoms on the plants.

5. Climate-Very high or low temperature, fog, hailstorm etc.
6. Sunlight-usually inadequate.
7. Insects-some kinds of sucking insects also produce disease like symptoms.

Due to non-parasitic diseases the plants become dull, remain dwarfed, the leaf margins may burn, the leaves may be blighted or develop clorotic or dead, dried spots of varying sizes on their surface, sometimes the leaves may become thick and rough and productivity of such plants is significantly reduced in terms of quantity as well as quality.

In severe cases the plants may succumb to these undesired factors. Neverthless as no parasitic microorganism are involved, There is no possibility of spread of these diseases and the trouble remains as localized one. Such diseases can be controlled if the cause is exactly known and remedial measures are possible and feasible.

FACTORS AFFECTING OCCURRENCES

Factors which affect Plant diseases are microorganisms, including fungi, bacteria, viruses, mycoplasmas, etc. or may be incited by physiological causes including high or low temperatures, lack or excess of soil moisture and aeration, deficiency or excess of plant nutrients, soil acidity or alkalinity, etc. Factors that limit the rate of disease development are the relatively low amounts of inoculum in the leg stage and the paucity of healthy plants available to the inoculum in the stationary stage.

The causative agents of disease in green plants number in a tens of thousands and include almost every form of life. But primary agents of disease may also be inanimate. Thus nonliving (abiotic) agents of disease include mineral deficiencies and excesses, air pollutants, biologically produced toxicants, improperly used pesticidal chemicals, and such other environmental factors as wind, water, temperature, and sunlight. Nonliving things certainly qualify as primary agents of disease; they continuously irritate plant cells and tissues; they are harmful to the physiological processes of the plant; and they evoke pathological responses that manifest as the symptoms characteristic of the several diseases. But the abiotic agents of disease in plants. The abiotic agents of plant disease are termed noninfectious, and the diseases they cause are termed noninfectious diseases.

Microorganisms: The microorganisms obtain their food either by breaking down dead plant and animal remains (saprophytes) or by attacking living plants and animals (parasites). In order to obtain nutrients, the parasitic organisms excrete enzymes or toxins and kill the cells of the tissues of the host plant, as a result of which either the whole plant or a part of it is damaged or killed, or considerable disturbance takes place in its normal metabolic processes.

Parasites: One of the factors causing plant diseases is parasites, those living organisms that can colonize the tissues of their host-plant victims and can be transmitted from plant to plant. These biotic agents are, therefore, infectious, and the diseases they cause are termed infectious diseases. The infectious agents of plant diseases are treated in the standard textbooks on plant pathology.

Ability to produce an inoculum: The parasitic pest must produce an inoculum, some structure that is adapted for transmission to a healthy plant and this can either parasitize the host directly or develop another structure that can establish a parasitic relationship with the host. For example, inocula for viruses are the viral particles (virions); for bacteria, the bacterial cells; for fungi, various kinds of spores or the hyphal threads of mold; for nematodes, eggs or second-stage larvae.

Agents/ Media for transportation of inoculum: The inoculum must be transported from its source to a part of a host plant that can be infected. This dispersal of inoculum to susceptible tissue is termed inoculation. Agents of inoculation may be insects (for most viruses and mycoplasmalike organisms and for some bacteria and fungi), wind (for many fungi), and splashing rain (for many fungi).

Wounds, Natural openings: The parasite must enter the host plant, which it can do (depending on the organism) in one or more of three ways; through wounds, through natural openings, or by growing directly through the unbroken protecting surface of the host. Viruses are literally injected into the plant as the homopterous insect carrier probes and feeds within its host. Bacteria depend on wounds or natural openings (for example, stomates, hydathodes, and lenticels) for entrance, but many fungi can penetrate plant parts by growing directly through plant surfaces, exerting enormous mechanical pressure and possibly softening host surfaces by enzymatic action.

Availability of food: For occurrence of disease one of the factor affecting is, availability of nourishment to grow within its host. This act of colonizations is termed infection. Certainly the parasite damages the cytomplasmic memberanes of the host cells, making those membranes freely permeable to solutes that would nourish the parasite And parasitism certainly results from enzymatic attacks by the parasite upon carbohydrates, proteins, and lipids inside the host cell. The breakdown products of such complex molecules would diffuse across the damaged host-cell membranes and be absorbed by the parasite in the form of sugars, amino acids, and the like. Air-borne parasites of foliage, flower, and fruit.

TRANSMISSION OF PLANT DISEASES

Rational and most effective control of parasitic plant diseases is possible only if –

1. the disease is correctly diagnosed,
2. the nature of transmission of the disease is known and
3. life cycle stages of the involved parasite i.e its mode of reproduction active structures produced under the favourable condition for repaid and wide dispersal and the structures produced to overcome adverse condition are known.

All parasitic as well as viral diseases are transmissible, the parasites or issues being infectious to suitable host plants with ability to spread from host to host and from one area to another. The microscopic parasites or sub-microscopic infectious agents viz. Viruses causing plant diseases are technically termed as 'pathogens'.

Wherever a disease is established in a particular area or country, transmission of the pathogen from host to host or from one place to another is termed as 'dissemination' or 'dispersal' of the pathogen. Dissemination of plant diseases is recognized in relation to different phases of diseases as under:

1. Primary infection: Contact of a pathogen with a suitable host plant and initiation of the disease first time in the season of a crop is called 'primary infection'. Often a few or several plants in the crop are likely to get primarily infected.
2. Secondary spread: When a plant or few plants are primarily infected, rapid multiplication of the pathogen sets in under favourable climatic conditions, which helps 'secondary spread' of the disease.

In case of fungal diseases the primary infections are usually caused by the resting or dormant structures of the fungal pathogen while the secondary spread occurs through the asexually produced, countless and short-lived spores

Although in majority of plant diseases the above two phases occur sequentially, primary infections occurring only once in the season and secondary spread often repeating several times in the same season thus causing rapid.

Transmission over wide areas, this type of cycle is not observed in many other disease. In such cases there is no secondary spread in the same season i.e. the pathogen multiplies only once during the crop period and the spread of disease is observed only during the following crop season by way of increased primary infection. Mildews, leafspots, blasts, blights etc are the common examples of former type while some smut disease of cereal or grain crops where black powder is formed in place of grains or inflorescence fall in the latter type. Control measures, therefore, in the former type need to be directed to avoid the primary infection and also to check the secondary spread.

Modes of primary infections: According to the general mode of primary infection plant diseases are recognized as:-

1. Soil borne
2. Seed borne, including diseases carried with planting material.
3. Wind borne
4. Insect borne etc.

This knowledge is helpful in adopting suitable control measures.

Continuous and discontinuous transmission: Transmission of disease is termed as 'continuous' when it occurs naturally by way of growth, multiplication and spread of the pathogen in an area or country where the disease is established.

At times, however, in an area or country where a particular disease has never occurred, it may get introduced through the agency of man carrying diseased material to a new locality or to a distant country for the purpose of introduction of new plants, crops, varieties etc. such transmission, of course, is unnatural and regarded as 'discontinuous' transmission. Altogether different control measures viz. Quarantine Regulations are enforced to check such discontinuous transmission.

Direct and indirect transmission: For classifying the methods of disease transmission in relation to the methods of suitable control measures, the following two groups can be conveniently recognized.

1. Direct transmission:-Disease transmission where the pathogen is carried externally or internally on the seed or planting material like cuttings, sets, tubers, bulbs etc.
2. Indirect transmission:-The pathogen spreading itself by way of its persistent growth or certain structures of the pathogen carried independently by natural agencies like wind, water, animals, insects, mites, nematodes, birds etc. are the different methods of indirect transmissions.

Direct transmission:-

1. Internal transmission through seed or planting material:-False smut disease as well as Helminthosporin Blight disease of wheat are the common examples of fungal diseases carried internally through apparently healthy seed. Ring rot and Brown rot of potato caused by bacteria are carried internally through the tubers. The well known whip smut and red rot of sugarcane are fungal diseases carried internally in the planting sets. Mosaic and leaf roll of potato which are viral diseases are also carried inside the infected tubers.
2. External transmission through seed or planting material:-In this mode of transmission the pathogen is carried externally over the surface of seed or vegetatively propagated plant parts like sets, tubers, bulbs etc. or may even be carried as a physical mixture of fungal structures with the seed. The common grain smut of jowar is an example of the former type while the fungal structures called 'sclerotia' having the size of a grain or slightly bigger in case of the Ergot disease of bajra are often likely to be transmitted in the form of physical mixture with the seed.

Indirect transmission:-

1. Autonomous transmission:-It takes place by continuous and persistent growth of the threads or 'hyphae' of the causal fungi in soil, characteristic of several wood rotting fungi attacking forest trees and some fruit plants. Some root rotting fungi infecting certain seasonal crops also are transmitted by this method. The autonomous dispersal of such soil fungi may range from few cm. To several (8 to 10) meters in a single season. Some plant parasitic nematodes also exhibit active but limited mobility in the soil.
2. Wind dispersal:-Fungal spores produced externally on host surfaces are most easily carried by wind currents and this is the most dangerous mode of transmission of plant pathogenic fungi like those causing powdery and downy mildews, leaf spots, blasts, blights and rust diseases. The black stem rust disease of wheat in India perpetuates on wild grasses in the Nilgiri hills in the south India from where the rust spores are carried to south, central & then to north India by wind currents every year. Spores may be carried from low to very high altitudes of 12,000 to 14,000 feet and from short distances to very long distances of several hundred kilometres.

   Extensive and severe epidemics of plant diseases are mostly the results of wind transmission of the pathogens. Wind dissemination involves four stages relating to the spores viz. Production of countless spores, their liberation in the wind currents, dispersal alongwith the wind and deposition on new susceptible host surfaces where they cause infection under favourable climatic conditions. Apart from spores, bits of fungal threads and nematode cysts are also amenable to wind transmission in certain cases.

3. Water dissemination: Disease transmission through the agency of water in different ways is comparatively less important as compared to the wind transmission. Splashing rain drops mostly transmit the foliar diseases from leaf to leaf, from shoot to shoot and even from plant to plant in case of closely spaced crops. Such transmission is usually accompanied by wind dispersal as well. Plant pathogens requiring high humidity conditions like the fungi causing downy mildew diseases or bacteria causing canker of citrus are well adapted to this kind of short distance water dispersal.

   Certain soil inhabiting pathogenic fungi and bacteria causing root and collar rots, wilts, foot, rots, etc are likely to be transmitted to much longer distances through the agencies like irrigation water, streams and rivers, etc. It is also an important agency in transmission of seeds of higher flowering parasites like dodder and striga.

4. Animals: Farm animals serve as disease transmitting agents in some cases. They are likely to carry the pathogen externally on their body surface, particularly on legs and hoofs, etc. or internally through their intestinal tract. Commonly, the soil inhabiting fungi causing rots and wilts are carried externally while certain smut fungi causing diseases to grain crops are transmitted through the intestinal tract.

5. Birds: Although birds play a very minor role in disease transmission, in cases of dispersal of seeds of higher flowering parasite. Loranthus sp. Parasitising certain trees like mango, etc. their role is of great significance. They transmit loranthus both externally and internally.

6. Implements and Tools: Farm implements used for cultivation of soil are often likely to transmit plant pathogens from one place to another. The pathogens in this case are usually carried in the form of bits of plant disease debris lying in the soil. Similarly tools used for carrying out operations like cutting, pruning, budding, grafting, thinning,etc. also help in the transmission of certain diseases from plant to plant. Several viral diseases are disseminated through the budding and grafting operations.

7. Insects: Most of the viral diseases of plants are transmitted through the agency of different insects. Both types of insects viz. sucking and chewing or/biting are capable of transmitting viral diseases. The transmission may be simply 'mechanical' or it may be 'biological'. In

the latter case the specific insect and the specific viral pathogen have some kind of association or relationship between the two. Insects in such cases are called the 'vectors' for the particular viral pathogen. In case of mechanical transmission the pathogen is simply carried externally or internally by the insect.

Viruses carried 'biologically' by the insect vectors are of two types:

1. Non-persistent-viral pathogen requiring no latent or incubation period in the insect body.
2. Persistent: viral pathogens requiring certain incubation period inside the vector body before they are inoculated or transmitted to healthy host. The insects responsible for transmission of viral diseases belong to the species of aphids, jassids (leaf hoppers), white flies, mealy bugs, etc. Certain bacterial and several fungal pathogens are also known to be carried by insects.
    1. Mites: Mites in contrast to insects are wingless anthropods resembling ticks and having four pairs of legs and no antennae. It is suspected that some viral diseases of chillies, tomato, brinjal, etc. have vector relationship with mites.
    2. Nematodes: Nematodes have been observed to transmit viral, bacterial and fungal plant diseases. Nematodes feeding externally on host plant roots cause injuries to roots which become the avenues for entrance of fungal and bacterial pathogens infecting plant roots. The Fan-leaf virus of grapevine is a well known example of transmission through a species of nematodes.
    3. Biological transmission: Dodder which is higher flowering parasite is known to transmit certain viral diseases which remain 'persistent' in the dodder plant. The flowering parasite after acquiring the virus from infected plant does not show any symptom itself but remains capable of transmitting the virus to healthy hosts.
    4. Human dispersal: Man is often responsible for transmission of plant diseases in two ways viz.
        1. Workers handling seedlings, other planting material or fruits are likely to get personally in contact with plant pathogens like fungi or bacteria. While handling the diseased material and unknowingly and indirectly transmit the pathogens to healthy seedlings or plant parts through his contaminated hands. This is a kind of 'continuous' mode of transmission.
        2. The other or'discontinous' mode of transmission for which only man is responsible is the most efficient and equally

dangerous phenomenon of transmission of plant diseases between distant geographical areas often separated by physical barriers like oceans, mountains or deserts, etc. Such long distances transmission of a disease to an area or country hitherto free from the disease is usually accomplished by the transport of infected seed, nursery stock or timber, etc. Thus it is a kind of direct transmission through propagating material.

In the history of plant pathology this mode of transmission has often resulted in to some of the worst plant disease epidemics in new areas, the local host plant stock being not adapted to the pathogen and consequently remaining highly susceptible. Fungal diseases viz. Late blight of potato and downy mildew of grapes, Bunchy top of banana, which is a viral disease and Bacterial blight of paddy are the examples of some severe and important diseases introducing in India from other countries.Practically all the countries of the world have suffered from such introduction of new plant diseases through the agency of man. Concerned international authorities, therefore, are engaged in promoting studies in this respect and designing regulations to reduce the danger.

THE STORING OF FRUITS AND VEGETABLES

The principles involved in the storing of perishable products, as fruits and vegetables, differ with the different commodities. All the root-crops, and most fruits, need to be kept in a cool, moist, and uniform temperature if they are to be preserved a great length of time. Squashes, sweet-potatoes, and some other things need to be kept in an intermediate and what might be called a high temperature; and the atmosphere should be drier than for most other products. The low temperature has the effect of arresting decomposition and the work of fungi and bacteria. The moist atmosphere has the effect of preventing too great evaporation and the consequent shriveling.

In the storing of any commodity, it is very important that the product is in proper condition for keeping. Discard all specimens that are bruised or are likely to decay. Much of the decay of fruits and vegetables in storage is not the fault of the storage process, but is really the work of diseases with which the materials are infected before they are put into storage. For example, if potatoes and cabbages are affected with the rot, it is practically impossible to keep them any length of time.

Apples, winter pears, and all roots, should be kept at a temperature somewhat near the freezing point. It should not rise above 40° F. for best results. Apples can be kept even at one or two degrees below the freezing point if the temperature is uniform. Cellars in which there are heaters are likely to be too dry and the temperature too high. In such places it is well to keep fresh vegetables and fruits in tight receptacles, and pack the roots in sand or moss in order to prevent shriveling. In these places, apples usually keep better if headed up in barrels than if kept on racks or shelves. In moist and cool cellars, however, it is preferable for the home supply to place them on shelves, not piling them more than five or six inches deep, for then they can be sorted over as occasion requires. In case of fruits, be sure that the specimens are not over-ripe when placed in storage. If apples are allowed to lie in the sun for a few days before being packed, they will ripen so much that it is very difficult to keep them.

Cabbages should be kept at a low and uniform temperature, and water should be drained away from them. They are stored in many ways in the field, but success depends so much on the season, particular variety, ripeness, and the freedom from injuries by fungi and insects, that uniform results are rarely secured by any one method. The best results are to be expected when they can be kept in a house built for the purpose, in which the temperature is uniform and the air fairly moist. When stored out of doors, they are likely to freeze and thaw alternately; and if the water runs into the heads, mischief results. Sometimes they are easily stored by being piled into a conical heap on well-drained soil and covered with dry straw, and the straw covered with boards. It does not matter if they are frosted, provided they do not thaw out frequently. Sometimes cabbages are laid head down in a shallow furrow plowed in well-drained land, and over them is thrown straw, the stumps being allowed to project through the cover. It is only in winters of rather uniform temperature that good results are to be expected from such methods. These are some of the main considerations involved in the storing of such things as cabbage.

In the storing of all products, especially those which have soft and green matter, as cabbages, it is well to provide against the heating of the produce. If the things are buried out of doors, it is important to put on a very light cover at first so that the heat may escape. Cover them gradually as the cold weather comes on. This is important with all vegetables that are placed in pits, as potatoes, beets, and the like. If covered deeply at once, they are likely to heat and rot. All pits made out of doors should be on well-drained and preferably sandy land.

When vegetables are wanted at intervals during the winter from pits, it is well to make compartment pits, each compartment holding a wagon load or whatever quantity will be likely to be wanted at each time. These pits are sunk in well-drained land, and between each of the two pits is left a wall of earth about a foot thick. One pit can then be emptied in cold weather without interfering with the others.

An outside cellar is better than a house cellar in which there is a heater, but it is not so handy. If it is near the house, it need not be inconvenient, however. A house is usually healthier if the cellar is not used for storage. House cellars used for storage should have a ventilating shaft.

Some of the principles involved in an ice-cooled storage house are explained. If the reader desires to make a careful study of storage and storage structures, he should consult cyclopedias and special articles.

The Forcing of Plants

There are three general means (aside from greenhouses) of forcing plants ahead of their season in the early spring—by means of forcing-hills and hand-boxes, by coldframes, and by hotbeds. The forcing-hill is an arrangement by means of which a single plant or a single "hill" of plants may be forced where it permanently stands. This type of forcing may be applied to perennial plants, as rhubarb and asparagus, or to annuals, as melons and cucumbers. A box with four removable sides, two of which are shown in end section in the figure, is placed around the rhubarb plant in the fall. The inside of the box is filled with straw or litter, and the outside is banked thoroughly with any refuse, to prevent the ground from freezing. When it is desired to start the plants, the covering is removed from both the inside and outside of the box and hot manure is piled around the box to its top.

If the weather is yet cold, dry light leaves or straw may be placed inside the box; or a pane or sash of glass may be placed on top of the box, when it will become a coldframe. Rhubarb, asparagus, sea-kale, and similar plants may be advanced two or four weeks by means of this method of forcing. Some gardeners use old barrels or half-barrels in place of the box. The box, however, is better and handier, and the sides can be stored for future use. Plants that require a long season in which to mature, and which do not transplant readily, as melons and cucumbers, may be planted in forcing-hills in the field.

A forcing-hill is sometimes made by digging a hole in the ground and planting the seeds in the bottom of it, placing the pane of glass upon a slight ridge or mound which is made on the surface of the ground. This method is less desirable than the other, because the seeds are placed in the poorest and coldest soil, and the hole is very likely to fill with water in the early days of spring. An excellent type of forcing-hill is made by the use of the hand-box. This is a rectangular box, without top or bottom, and a pane of glass is slipped into a groove at the top. It is really a miniature coldframe. The earth is banked up slightly about the box, in order to hold it against winds and to prevent the water from running into it. If these boxes are made of good lumber and painted, they will last for many years. Any size of glass may be used which is desired, but a ten-by-twelve pane is as good as any for general purposes. After the plants are thoroughly established in these forcing-hills, and the weather is settled, the protection is wholly removed, and the plants grow normally in the open.

A very good temporary protection may be given to tender plants by using four panes of glass, the two inner panes being held together at the top by a block of wood through which four nails are driven. Plants are more likely to burn in these glass frames than in the hand-boxes, and such frames are not so well adapted to the protection of plants in very early spring; but they are often useful for special purposes.

In all forcing-hills, as in coldframes and hotbeds, it is exceedingly important that the plants receive plenty of air on bright days. Plants that are kept too close become weak or "drawn", and lose the ability to withstand changes of weather when the protection is removed. Even though the wind is cold and raw, the plants inside the frames ordinarily will not suffer if the glass is taken off when the sun is shining.

Coldframes

A coldframe is nothing more than an enlarged hand-box; that is, instead of protecting but a single plant or a single hill with a single pane of glass, the frame is covered with sash, and is large enough to accommodate many plants.

There are three general purposes for which a coldframe is used: For the starting of plants early in spring; for receiving partially hardened plants that have been started earlier in hotbeds and forcing-houses; for wintering young cabbages, lettuce, and other hardy plants that are sown in the fall.

Coldframes are ordinarily placed near the buildings, and the plants are transplanted into the field when settled weather comes. Sometimes, however, they are made directly in the field where the plants are to remain, and the frames, and not the plants, are removed. When used for this latter purpose, the frames are made very cheap by running two rows of parallel planks through the field at a distance apart of six feet. The plank on the north is ordinarily ten to twelve inches wide, and that on the south eight to ten inches. These planks are held in place by stakes, and the sashes are laid across them. Seeds of radishes, beets, lettuce, and the like, are then sown beneath the sash, and when settled weather arrives, the sash and planks are removed and the plants are growing naturally in the field. Half-hardy plants, as those mentioned, may be started fully two or three weeks in advance of the normal season by this means.

One of the simplest types of coldframes is a lean-to against the foundation of a house. A sill is run just above the surface of the ground, and the sashes are laid on rafters which run from this sill to the sill of the house. If this frame is on the south side of the building, plants may be started even as early as a month before the opening of the season. Such lean-to frames are sometimes made against greenhouses or warm cellars, and heat is supplied to them by the opening of a door in the wall. In frames that are in such sunny positions as these, it is exceedingly important that care be taken to remove the sash, or at least to give ample ventilation, in all sunny days.

A different type of lean-to structure may be either a temporary or permanent building, and it is generally used for the protection of half-hardy plants that are grown in pots and tubs. It may be used, however, for the purpose of forwarding pot-plants early in the spring and for protection of peaches, grapes, oranges, or other fruits in tubs or boxes. If it is desired merely to protect the plants through the winter, it is best to have the structure on the north side of the building, in order that the sun may not force the plants into activity. Another structure that may be used both to carry half-hardy plants over winter and for starting plants early in spring is really a miniature greenhouse without heat. It is well adapted for mild climates.

The common type of coldframe is twelve feet long and six feet wide, and is covered with four three-by-six sash. It is made of ordinary lumber loosely nailed together. If one expects to use coldframes or hotbeds every year, however, it is advisable to make the frames of two-inch stuff, well painted, and to join the parts by bolts and tenons, so that they may be taken apart and stored until needed for the next year's crop.

It is always advisable to place coldframes and hotbeds in a protected place, and particularly to protect them from cold north winds. Buildings afford excellent protection, but the sun is sometimes too hot on the south side of large and light-coloured buildings. One of the best means of protection is to plant a hedge of evergreens. It is always desirable, also to place all the coldframes and hotbeds close together, for the purpose of economizing time and labour. A regular area or yard may be set aside for this purpose. Various small and portable coldframes may be used about the garden for the protection of tender plants or to start them early in the spring. Pansies, daisies, and border carnations, for example, may be brought on very early by setting such frames over them or by planting them under the frames in the fall. These frames may be of any size desired, and the sash may be either removable, or, in case of small frames, they may be hinged at the top.

Hotbeds

A hotbed differs from a coldframe in being provided with bottom heat. This heat is ordinarily supplied by means of fermenting manure, but it may be obtained from other fermenting material, as tanbark or leaves, or from artificial heat, as flues, steam pipes, or water pipes.

The hotbed is used for the very early starting of plants; and when the plants have outgrown the bed, or have become too thick, they are transplanted into cooler hotbeds or into coldframes. There are some crops, however, that are carried to full maturity in the hotbed itself, as radishes and lettuce.

The date at which the hotbed may be started with safety depends almost entirely on the means at command of heating it and on the skill of the operator. In the northern states, where outdoor gardening does not begin until the first or the last of May, hotbeds are sometimes started as early as January; but they are ordinarily delayed until early in March. The heat for hotbeds is commonly supplied by the fermentation of horse manure. It is important that the manure be as uniform as possible in composition and texture, that it come from highly fed horses, and is practically of the same age. The best results are usually secured with manure from livery stables, from which it can be obtained in large quantities in a short space of time. Perhaps as much as one half of the whole material should be of litter or straw that has been used in the bedding.

The manure is placed in a long and shallow square-topped pile, not more than four or six feet high, as a rule, and is then allowed to ferment. Better results are generally obtained if the manure is piled under cover. If the weather is cold and fermentation does not start readily, wetting the pile with hot water may start it. The first fermentation is nearly always irregular; that is, it begins unequally in several places in the pile. In order to make the fermentation uniform, the pile must be turned occasionally, taking care to break up all hard lumps and to distribute the hot manure throughout the mass. It is sometimes necessary to turn the pile five or six times before it is finally used, although half this number of turnings is ordinarily sufficient. When the pile is steaming uniformly throughout, it is placed in the hotbed, and is covered with the earth in which the plants are to be grown.

Hotbed frames are sometimes set on top of the pile of fermenting manure. The manure should extend some distance beyond the edges of the frame; otherwise the frame will become too cold about the outside, and the plants will suffer. It is preferable, however, to have a pit beneath the frame in which the manure is placed.

If the bed is to be started in midwinter or very early in the spring, it is advisable to make this pit in the fall and to fill it with straw or other litter to prevent the earth from freezing deep. When it is time to make the bed, the litter is thrown out, and the ground is warm and ready to receive the fermenting manure.

The pit should be a foot wider on either side than the width of the frame. Upon the ground a layer of an inch or two of any coarse material is placed to keep the manure off the cold earth. Upon this, from twelve to thirty inches of manure is placed. Above the manure is a thin layer of leafmold or some porous material, that will serve as a distributor of the heat, and above this is four or five inches of soft garden loam, in which the plants are to be grown. It is advisable to place the manure in the pit in layers, each stratum to be thoroughly trodden down before another one is put in. These layers should be four to eight inches in thickness.

By this means the mass is easily made uniform in consistency. Manure that has too much straw for the best results, and which will therefore soon part with its heat, will spring up quickly when the pressure of the feet is removed. Manure that has too little straw, and which therefore will not heat well or will spend its heat quickly, will pack down into a soggy mass underneath the feet. When the manure has sufficient litter, it will give a springy feeling to the feet as a person walks over it, but will not fluff up when the pressure is removed. The quantity of manure to be used will depend on its quality, and also on the season in which the hotbed is made. The earlier the bed is made, the larger should be the quantity of manure. Hotbeds that are intended to hold for two months should have about two feet of manure, as a rule.

The manure will ordinarily heat very vigorously for a few days after it is placed in the bed. A soil thermometer should be thrust through the earth down to the manure, and the frame kept tightly closed. When the temperature is passing below 90°, seeds of the warm plants, like tomatoes, may be sown, and when it passes below 80° or 70°, the seeds of cooler plants may be sown.

If hotbeds are to be used every year, permanent pits should be provided for them. Pits are made from two to three feet deep, preferably the former depth, and are walled up with stone or brick. It is important that they be given good drainage from below. In the summer-time, after the sash are stripped, the old beds may be used for the growing of various delicate crops, as melons or half-hardy flowers. In this position, the plants can be protected in the fall. As already suggested, the pits should be cleaned out in the fall and filled with litter to facilitate the work of making the new bed in the winter or spring. Various modifications of the common type of hotbed will suggest themselves to the operator.

The frames should ordinarily run in parallel rows, so that a man walking between them can attend to the ventilation of two rows of sash at once. There are two parallel runs, with walks on the outside, and between them are racks to receive the sash from the adjacent frames. The sash from the left-hand bed are run to the right, and those from the right-hand bed are run to the left. Running on racks, the operator does not need to handle them, and the breakage of glass is therefore less; but this system is little used because of the difficulty of reaching the farther side of the bed from the single walk. If the hotbed were high enough and broad enough to allow a man to work inside, we should have a forcing-house. Such a structure has one side of which the manure and soil are already in place. These manure-heated houses are often very efficient, and are a good make-shift until such time as the gardener can afford to put in flue or pipe heat. Hotbeds may be heated by means of steam or hot water. They can be piped from the heater in a dwelling-house or greenhouse.

It would scarcely pay to put in a hot water or steam heater for the express purpose of heating hotbeds, for if such an expense were incurred, it would be better to make a forcing-house. Hotbeds may be heated, however, with hot-air flues with very good results. A home-made brick furnace may be constructed in a pit at one end of the run and underneath a shed, and the smoke and hot air, instead of being carried directly upwards, is carried through a slightly rising horizontal pipe that runs underneath the beds.

For some distance from the furnace, this flue may be made of brick or unvitrified sewer pipe, but stove-pipe may be used for the greater part of the run. The chimney is ordinarily at the farther end of the run of beds.

It should be high, in order to provide a good draft. If the run of beds is long, there should be a rise in the underlying pipe of at least one foot in twenty-five. The greater the rise in this pipe, the more perfect will be the draft. If the runs are not too long, the underlying pipe may return underneath the beds and enter a chimney directly over the back end of the furnace, and such a chimney, being warmed from the furnace, will ordinarily have an excellent draft. The underlying pipe should occupy a free space or pit beneath the beds, and whenever it lies near to the floor of the bed or is very hot, it should be covered with asbestos cloth. While such flue-heated hotbeds may be eminently successful with a grower or builder of experience, it may nevertheless be said, as a general statement, that whenever such trouble and expense are incurred, it is better to make a forcing-house. The subject of forcing-houses and greenhouses is not discussed in this book. The most satisfactory material for use in hotbed and cold-frame sash is double-thick, second-quality glass; and panes twelve inches wide are ordinarily broad enough, and they suffer comparatively little in breakage. For coldframes, however, various oiled papers and waterproof cloths may be used, particularly for plants that are started little in advance of the opening of the season. When these materials are used, it is not necessary to have expensive sash, but rectangular frames are made from strips of pine seven-eighths inch thick and two and one-half inches wide, halved together at the corners and each corner reënforced by a square carriage-corner, such as is used by carriage-makers to secure the corners of buggy boxes. These corners can be bought by the pound at hardware stores.

Management of Hotbeds

Close attention is required in the management of hotbeds, to insure that they do not become too hot when the sun comes out suddenly, and to give plenty of fresh air. Ventilation is usually effected by raising the sash at the upper end and letting it rest upon a block. Whenever the temperature is above freezing point, it is generally advisable to take the sash off part way or even to strip it off entirely. Care should be taken not to water the plants at nightfall, especially in dull and cold weather, but to give them water in the morning, when the sun will soon bring the temperature up to its normal state. Skill and judgment in watering are of the greatest importance in the management of hotbeds; but this skill comes only from thoughtful practice. The satisfaction and effectiveness of the work are greatly increased by good hose connections and good watering-pots. Some protection, other than the glass, must be given to hotbeds. They need covering on every cold night, and sometimes during the entire day in very severe weather. Very good material for covering the sash is matting, such as is used for covering floors. Old pieces of carpet may also be used. Various hotbed mattings are sold by dealers in gardeners' supplies.

Gardeners often make mats of rye straw, although the price of good straw and the excellence of manufactured materials make this home-made matting less desirable than formerly. Such mats are thick and durable, and are rolled up in the morning. There are various methods of making these straw mats. A frame is made after the manner of a saw-horse, with a double top, and tarred or marline twine is used for securing the strands of straw. It is customary to use six runs of this warp. Twelve spools of string are provided, six hanging on either side. Some persons wind the cord upon two twenty-penny nails, as shown in the figure, these nails being held together at one end by wire which is secured in notches filed into them. The other ends of the spikes are free, and allow the string to be caught between them, thus preventing the balls from unwinding as they hang upon the frame. Two wisps of straight rye straw are secured and laid upon the frame, with the butt ends outward and the heads overlapping. Two opposite spools are then brought up, and a hard knot is tied at each point. The projecting butts of the straw are then cut off with a hatchet, and the mat is allowed to drop through to receive the next pair of wisps. In making these mats, it is essential that the rye contains no ripe grain; otherwise it attracts the mice. It is best to grow rye for this especial purpose, and to cut it before the grain is in the milk, so that the straw does not need to be threshed.

In addition to these coverings of straw or matting, it is sometimes necessary to provide board shutters to protect the beds, particularly if the plants are started very early in the season. These shutters are made of half-inch or five-eighths-inch pine lumber, and are the same size as the sash—three by six feet. They may be placed upon the sash underneath the matting, or they may be used above the matting. In some cases they are used without any matting.

In the growing of plants in hotbeds, every effort should be made to prevent the plants from growing spindling, or becoming "drawn." To make stocky plants, it is necessary to give room to each plant, to be sure that the distance from the plants to the glass is not great, to provide not too much water in dull and cold weather, and particularly to give abundance of air.

2

Ornamental Plants

An ornamental plant is a plant that is grown for its ornamental qualities, rather than for its commercial or other value. The term is often abbreviated to ornamental (usually as a noun) when used in horticultural contexts.

Ornamental plants are typically grown in the flower garden, or as house plants. Most commonly they are grown for display of their flowers. Other common ornamental features include leaves, scent, fruit, stem and bark. In some cases, unusual features may be considered ornamental, such as the prominent and rather vicious thorns of *Rosa sericea*. In all cases, their purpose is the enjoyment of gardeners and visitors. Ornamental plants may also be used for landscaping, and for cut flowers.

Similarly trees may be called ornamental trees. This term is used when they are used as part of a garden setting, for instance for their flowers, their shapes or for other attractive characteristics. By comparison, trees used in larger landscape effects such as screening and shading, or in urban and roadside plantings, are called *amenity trees.*

For plants to be considered as ornamental, they may require specific work and activity by a gardener. For instance, many plants cultivated for topiary and bonsai would only be considered as ornamental by virtue of the regular pruning carried out on them by the gardener, and they may rapidly cease to be ornamental if the work was abandoned.

Ornamental plants and trees are distinguished from utilitarian and crop plants, such as those used for agriculture and vegetable crops, and for forestry or as fruit trees. This does not preclude any particular type of plant being grown both for ornamental qualities in the garden, and for utilitarian purposes in other settings. Thus lavender is typically grown as an ornamental plant in gardens, but may also be grown as a crop plant for the production of lavender oil.

SOME IMPORTANT ORNAMENTAL PLANTS

Abelia

Abelia is a genus of about 15-30 species and many hybrids in the honeysuckle family Caprifoliaceae, in the part of that family split off by some authors in the segregate family Linnaeaceae. The Angiosperm Phylogeny Group considers Linnaeaceae to encompass such genera as Linnaea, Abelia, Dipelta, Kolkwitzia, and Zabelia.

Abelias are shrubs from 1-6 m tall, native to eastern Asia (Japan west to the Himalaya) and southern North America (Mexico); the species from warm climates are evergreen, and colder climate species deciduous. The leaves are opposite or in whorls of three, ovate, glossy, dark green, 1.5-8 cm long, turning purplish-bronze to red in autumn in the deciduous species. The flowers appear in the upper leaf axils and stem ends, 1-8 together in a short cyme; they are pendulous, white to pink, bell-shaped with a five-lobed corolla, 1-5 cm long, and usually scented. Flowering continues over a long and continuous late spring to fall period.

Abelias are popular garden shrubs. The most widely grown is the hybrid *Abelia x grandiflora* (Glossy Abelia; hybrid *Abelia chinensis* x *Abelia uniflora*). This is a rounded, spreading, multi-stemmed shrub with gracefully arching branches to 1-1.8 m tall, with ovate, glossy, dark green semi-evergreen leaves to 2-6 cm long, and clusters of white-tinged-pink, bell-shaped flowers to 2 cm long.

Abelia species are used as food plants by the larvae of some Lepidoptera species - see list of Lepidoptera which feed on Abelia.

Abelmoschus

Abelmoschus is a genus of about fifteen species of flowering plants in the mallow family Malvaceae, native to tropical Africa, Asia and northern Australia. It was formerly included within *Hibiscus*, but is now classified as a distinct genus.

The genus comprises annual and perennial herbaceous plants, growing to 2 m tall. The leaves are 10-40 cm long and broad, palmately lobed with 3-7 lobes, the lobes very variable in depth, from barely lobed, to cut almost to the base of the leaf. The flowers are 4-8 cm diameter, with five white to yellow petals, often with a red or purple spot at the base of each petal. The fruit is a capsule, 5-20 cm long, containing numerous seeds. Abelmoschus species are used as food plants by the larvae of some Lepidoptera species including *Chionodes hibiscella* which has been recorded on *A. moschatus*.

Selected Species

- Abelmoschus caillei - (syn. Hibsicus manihot var. caillei). West African okra
- Abelmoschus esculentus - (syn. Hibiscus esculentus). Okra
- Abelmoschus manihot - (syn. Hibiscus manihot). Aibika
- Abelmoschus moschatus - (syn. Hibiscus abelmoschus). Musk Mallow

Several species are edible, with both the young seed pods and the young leaves being eaten as a vegetable. The most important commercially-grown species is Okra.

Abelmoschus manihot (aibika) furnishes cordage like jute, and Abelmoschus moschatus (musk mallow) is grown for musk seeds (musk ambrette, which causes photoallergy).

Acaena

Acaena is a genus of about 100 species of perennial herbs and subshrubs in the Rosaceae, native mainly to the Southern Hemisphere, notably New Zealand, Australia and South America, but with a few species extending into the Northern Hemisphere, north to Hawaii (*A. exigua*) and California (A. pinnatifida).

The leaves are alternate, 4-15 cm long, and pinnate or nearly so, with 7-21 leaflets. The flowers are produced in a tight globose inflorescence 1-2 cm diameter, with no petals. The fruit is also a dense ball of many seeds; in many (but not all) species the seeds bear a barbed arrowhead point, the seedhead forming a burr which attaches itself to animal fur or feathers for dispersal.

Firs (Abies)

Firs (*Abies*) are a genus of between 45-55 species of evergreen conifers in the family Pinaceae. All are trees, reaching heights of 10-80 m (30-260 ft) tall and trunk diameters of 0.5-4 m (2-12 ft) when mature. Firs can be distinguished from other members of the pine family by their needle-like leaves, attached to the twig by a base that resembles a small suction cup; and by erect, cylindrical cones 5-25 cm (2-10 in) long that disintegrate at maturity to release the winged seeds. Identification of the species is based on the size and arrangement of the leaves, the size and shape of the cones, and whether the bract scales of the cones are long and exserted, or short and hidden inside the cone. They are most closely related to the cedars (*Cedrus*). Firs are found through much of North and Central America, Europe, Asia, and North Africa, occurring in mountains over most of the range.

Firs are used as food plants by the larvae of some Lepidoptera species including *Chionodes abella* (recorded on White Fir), Autumnal Moth, Conifer Swift (a pest of Balsam Fir), The Engrailed, Grey Pug, Mottled Umber and Pine Beauty. Douglas-firs are not true firs, being of the genus *Pseudotsuga.*

The wood of most firs is considered unsuitable for general timber use, and is often used as pulp or for the manufacture of plywood and rough timber. Nordmann Fir, Noble Fir, Fraser Fir and Balsam Fir are very popular Christmas trees, generally considered to be the best trees for this purpose, with aromatic foliage that does not shed many needles on drying out. Many are also very decorative garden trees, notably Korean Fir and Fraser Fir, which produce brightly coloured cones even when very young, still only 1-2 m (3-6 ft) tall.

Acorus

Acorus is a genus of monocot flowering plants. This genus was once placed within the family Araceae (aroids), but more recent phylogenies place it in its own family Acoraceae and order Acorales, of which it is the sole genus of the oldest surviving line of monocots. The exact relationship of Acorus to other monocots, however, is still debated by scientitst. Some studies indicate that it is placed in a lineage (the order Alismatales), that also includes aroids (Araceae), Tofieldiaceae, and several families of aquatic monocots (e.g., Alismataceae, Posidoniaceae). Common names include Calamus and Sweet Flag. The name 'acorus' is derived from the Greek word 'acoron', a name used by Dioscorides, which in turn was derived from 'coreon', meaning 'pupil', because it was used in herbal medicine as a treatment for inflammation of the eye.

The genus is native to North America and northern and eastern Asia, and naturalised in southern Asia and Europe from ancient cultivation. The known wild populations are diploid except for some tetraploids in eastern Asia, while the cultivated plants are sterile triploids, probably of hybrid origin between the diploid and tetraploid forms.

Aerangis

Aerangis is a genus of the Orchid family (Orchidaceae) (subfamily Vandoideae, tribe Vandeae, subtribe Aerangidinae).

The name of this genus has been derived from the Greek words 'aer' (air) and 'angos' (urn), referring to the form of the lip.

Approximately 50 species in this genus are known mostly from tropical Africa, but also from the Comoros Islands, Madagascar and Sri Lanka. These include:

- Aerangis curnowiana : Pearl of Madagascar
- Aerangis punctata : Pearl of Madagascar
- Aerangis luteoalba var. rhodosticta: Pheasant-eye Orchid

Species are usually epiphytic, sometimes lithophytic small orchids, resembling *Vanda*s in appearance. Their large, waxy, star-shaped flowers are generally white, cream-colored or yellow. They show a long, nectar-filled spur, often longer than the flower itself. There is a single stem with many flowers on a long raceme. The lip is flat and resembles the petals and sepals. They give off an agreeable smell during the night. There are six to ten parallel-veined, fleshy, evergreen leaves. These orchids are not often found in collections, even though they are rather easy to grow.

Wild Ginger

Wild ginger refers to an herbaceous plant genus *Asarum* of the birthwort family Aristolochiaceae.

Asarum canadense is native to the forests of eastern North America. It is found from the Great Plains east to the Atlantic Coast, and from southeastern Canada south to approximately the fall line in the southeastern United States.

Asarum caudatum is found in British Columbia south through Washington and Oregon to central California, and from the Coast Range east to western Montana.

The plant is called wild ginger because the rhizome tastes and smells similar to that of ginger root, but the two are not particularly related. The root can be used as a spice, but is a potent diuretic, or urinary stimulant. *Asarum canadense* and other species in the genus contain the chemical aristolochic acid, which is carcinogenic in rats. The birthwort family also contains the Aristolochia genus. Aristolochia is a human carcinogen.

Wild ginger favors moist, shaded sites with humus-rich soil. The deciduous, heart-shaped leaves are opposite, and borne from the rhizome which lies just under the soil surface. Two leaves emerge each year from the growing tip. The curious jug-shaped flowers, which give the plant an alternate name, little jug, are borne singly in Spring between the leaf bases. Wild ginger can easily be grown in a shade garden, and makes an attractive groundcover.

Aloe

Aloe, also written Aloë, is a genus containing about four hundred species of flowering succulent plants. The genus is native to Africa and is common in South Africa's Cape Province and the mountains of tropical Africa, and neighbouring areas such as Madagascar, the Arabian peninsula and the islands off Africa.

The APG II system (2003) placed the genus in the family Asphodelaceae. In the past it has also been assigned to families Aloaceae and Liliaceae. Members of the closely allied genera *Gasteria, Haworthia* and *Kniphofia* which have a similar mode of growth, are also popularly known as aloes. Note that the plant sometimes called "American aloe" (*Agave americana*), belongs to Agavaceae, a different family.

Most Aloes have a rosette of large, thick, fleshy leaves. The leaves are often lance-shaped with a sharp apex and a spiny margin. Aloe flowers are tubular, frequently yellow, orange or red and are borne on densely clustered, simple or branched leafless stems.

Many species of Aloe are seemingly stemless, with the rosette growing directly at ground level; other varieties may have a branched or un-branched stem from which the fleshy leaves spring. They vary in colour from grey to bright green and are sometimes striped or mottled.

Bacopa Monnieri

Bacopa monnieri is a perennial, creeping herb whose habitat includes wetlands and muddy shores. Common names include Water Hyssop and brahmi (note: brahmi is also the Ayurvedic name given to *Centella asiatica* and other herbs).

The leaves of this plant are succulent and relatively thick. Leaves are oblanceolate and are arranged oppositely on the stem. The flowers are small and white, with four or five petals. Its ability to grow in water makes it a popular aquarium plant. It can even grow in slightly brackish conditions. Propagation is often achieved through cuttings.

It commonly grows in marshy areas throughout India, Nepal, Sri Lanka, China, Taiwan, and is also found in Florida and other southern states of USA where it can be grown in damp conditions by the pond or bog garden.

Famed in Ayurvedic medicine, brahmi has antioxidant properties. It has been reported to reduce oxidation of fats in the blood stream, which is a risk factor for cardiovascular diseases.

It has been used for centuries to help benefit epilepsy, memory capacity, increase concentration, and reduce stress-induced anxiety. It is listed as a nootropic, a drug that enhances cognitive ability.

In India, this plant has also been used traditionally to consecrate new born babies in the belief that it will open the gateway of intelligence. Recent studies suggest bacopa may improve intellectual activity. This plant is also known as Thyme-leafed gratiola and Moneywort. Synonyms include Herpestris monnieria, Moniera euneifolia, Lysimachia monnieri, and Bacopa Monniera.

Begonia

Begonia is a genus in the flowering plant family Begoniaceae. The only other member of the family Begoniaceae is *Hillebrandia,* a genus with a single species in the Hawaiian Islands. The genus *Symbegonia* is now included in *Begonia.* "Begonia" is the common name as well as the generic name for all members of the genus.

With 1500+ species, *Begonia* is one of the ten largest angiosperm genera. The species are terrestrial (sometimes epiphytic) herbs or undershrubs and occur in subtropical and tropical moist climates, in South and Central America, Africa and southern Asia. Terrestrial species in the wild are commonly upright-stemmed, rhizomatous, or tuberous. The plants are monoecious, with unisexual male and female flowers occurring separately on the same plant, the male containing numerous stamens, the female having a large inferior ovary and two to four branched or twisted stigmas. In most species the fruit is a winged capsule containing numerous minute seeds, although baccate fruits are also known. The leaves, which are often large and variously marked or variegated, are usually asymmetric (unequal-sided).

Because of their sometimes showy flowers of white, pink, scarlet or yellow color and often attractively marked leaves, many species and innumerable hybrids and cultivars are cultivated. The genus is unusual in that species throughout the genus, even those coming from different continents, can frequently be hybridized with each other, and this has led to an enormous number of cultivars. The American Begonia Society classifies begonias into several major groups: cane-like, shrub-like, tuberous, rhizomatous, semperflorens, rex, trailing-scandent, or thick-stemmed. For the most part these groups do not correspond to any formal taxonomic groupings or phylogeny and many species and hybrids have characteristics of more than one group, or fit well into none of them. The genus name honors Michel Bégon, a French patron of botany.

The different groups of begonias have different cultural requirements but most species come from tropical regions and therefore they and their hybrids require warm temperatures. Most are forest understory plants and require bright shade; few will tolerate full sun, especially in warmer climates. In general, begonias require a well-drained growing medium that is neither constantly wet nor allowed to dry out completely. Many begonias will grow and flower year-round but tuberous begonias usually have a dormant period, during which the tubers can be stored in a cool and dry place.

Begonias of the semperflorens group are frequently grown as bedding plants outdoors. A recent group of hybrids derived from this group is marketed as "Dragonwing Begonias"; they are much larger both in leaf and in flower. Tuberous begonias are frequently used as container plants. Although most *Begonia* species are tropical or subtropical in origin, the Chinese species *B. grandis* is hardy to USDA hardiness zone 6 and is commonly known as the "hardy begonia". Most begonias can be grown outdoors year-round in subtropical or tropical climates, but in temperate climates begonias are grown outdoors as annuals, or as house or greenhouse plants.

Most begonias are easily propagated by division or from stem cuttings. In addition, many can be propagated from leaf cuttings or even sections of leaves, particularly the members of the rhizomatous and rex groups.

Bougainvillea

Bougainvillea is a genus of flowering plants native to the tropical and subtropical regions of Asia from Pakistan to Afghanistan, and South America from Brazil west to Peru and south to southern Argentina (Chubut Province). Different authors accept between four and 18 species in the genus. The name comes from Louis Antoine de Bougainville, an admiral in the French Navy who discovered the plant in Brazil in 1768.

They are thorny, woody, vines growing anywhere from 1-12 meters tall, scrambling over other plants with their hooked thorns. They are evergreen where rainfall occurs all year, or deciduous if there is a dry season. The leaves are alternate, simple ovate-acuminate, 4-13 cm long and 2-6 cm broad. The actual flower of the plant is small and generally white, but each cluster of three flowers is surrounded by three or six bracts with the bright colors associated with the plant, including pink, magenta, purple, red, orange, white, or yellow. *Bougainvillea glabra* is sometimes referred to as "paper flower" because the bracts are thin and papery. The fruit is a narrow five-lobed achene.

Bougainvillea are relatively pest-free plants, but may suffer from worms and aphids. The larvae of some Lepidoptera species also use them as food plants including Giant Leopard Moth.

Bougainvilleas are popular ornamental plants in most areas with warm climates, including India, Taiwan, Vietnam, Malaysia, Australia, the Mediterranean region, the Caribbean, Mexico, South Africa, and the United States in Arizona, California, Florida, Hawaii, and southern Texas.

Numerous cultivars and hybrids have been selected, including nearly thornless shrubs. Some Bougainvillea cultivars are sterile, and are propagated from cuttings.

Bougainvillea are rapid growing and flower all year in warm climates, especially when pinched or pruned. Bloom cycles are typically four to six weeks. Bougainvillea grow best in very bright full sun and with frequent fertilization, but the plant requires little water to flower. As indoor houseplants in temperate regions, they can be kept small by bonsai techniques. If overwatered, Bougainvillea will not flower and may lose leaves or wilt, or even die from root decay.

Berberis

Berberis (Pepperidge bush, Pepperidge-bushis) a genus of about 450-500 species of deciduous and evergreen shrubs from 1-5 m tall with thorny shoots, native to the temperate and subtropical regions of Europe, Asia, Africa, North America and South America. They are closely related to the genus Mahonia, which is included within Berberis by some botanists. Many are known by the vernacular name barberry.

The genus is characterised by dimorphic shoots, with long shoots which form the structure of the plant, and short shoots only 1-2 mm long. The leaves on long shoots are non-photosynthetic, developed into three-spined thorns 3-30 mm long; the bud in the axil of each thorn-leaf then develops a short shoot with several normal, photosynthetic leaves. These leaves are 1-10 cm long, simple, and either entire, or with spiny margins. Only on young seedlings do leaves develop on the long shoots, with the adult foliage style developing after the young plant is 1-2 years old.

The deciduous species (e.g. Berberis thunbergii, B. vulgaris) are noted for good autumn colour, the leaves turning pink or red before falling. In some evergreen species from China (e.g. B. candidula, B. verruculosa), the leaves are brilliant white beneath, making them particularly attractive.

The flowers are produced singly or in racemes of up to 20 on a single flower-head. They are yellow or orange, 3-6 mm long, with six sepals and six petals in alternating whorls of three, the sepals usually coloured like the petals. The fruit is a small berry 5-15 mm long, ripening red or dark blue, often with a pink or violet waxy surface bloom; they may be either long and narrow (like a bar, hence 'barberry') or in other species, spherical.

The berries are edible, and rich in vitamin C, though with a very sharp flavour; the thorny shrubs make harvesting them difficult, so in most places they are not widely consumed. They are an important food for many small birds, which disperse the seeds in their droppings. In Iran, the dried fruits, known as zereshk, are more widely used, imparting a tart flavor to chicken dishes. In Ukraine, they are used as the flavouring in a popular candy of the same name.

Berberis species are used as food plants by the larvae of some Lepidoptera species including Mottled Pug.

Several are popular garden shrubs, grown for their ornamental leaves, yellow flowers, and red or blue-black berries. They are also valued for crime prevention; being very dense, viciously spiny shrubs, they make very effective barriers impenetrable to burglars.

For this reason they are often planted below potentially vulnerable windows, and used as hedges and other barriers.

Berberis buxifolia (Calafate) and Berberis darwinii (Michay) are two species found in Patagonia in Argentina and Chile. Their edible purple fruits are used for jams and infusions; anyone who tries a berry is said to be certain to return to Patagonia. The calafate and michay are symbols of Patagonia.

Berberis vulgaris (European Barberry) is the alternate host species of the wheat rust Puccinia graminis, a serious fungal disease of wheat. For this reason, cultivation of this species is prohibited in many areas.

Some Berberis have become invasive species when planted outside of their native ranges, including B. glaucocarpa and B. darwinii in New Zealand (where it is now banned from sale and propagation), and B. thunbergii in some parts of North America.

Calypso Orchid

The Calypso orchid (Calypso bulbosa), also known as the fairy slipper or Venus's slipper, is a small pink, purple, or red flowered (accented with white lower lip, darker purple spottings, and yellow beard) perennial member of the orchid family (Orchidaceae), and is found in undisturbed northern montane forests.

It is the only species currently classified in the genus Calypso, which takes its name from the Greek signifying concealment, as they tend to favor sheltered areas on conifer forest floors.

Their tiny purple blooms, typically about 10 cm in height, can be a pleasant sporadic sight on hiking trails from late March onwards, though in the more northerly parts of their range they do not bloom until May and June. These come to full bloom in nearly 20 years.

Its range is circumpolar, and includes all the western states and most of the most northerly states of the United States. Furthermore Scandinavia (northern Sweden and Finland), northern part of European Russia and eastern Siberia and Canada - see external links for map. Two varieties are found in the USA, var. *americana* and var. *occidentalis,* which are found respectively east and west of the Sierra Nevada ranges.

Although the calypso orchid's distribution is wide, it is very susceptible to disturbance, and is therefore classified as threatened or endangered in several states, and in Sweden and Finland as well. It is easily disturbed and does not transplant well, owing to its mycorrhizal dependence on specific soil fungi. The bulbs have been used as a food source by North American native peoples, though this is not recommended now because the sites for these plants are now rare and easily destroyed. The Thompson Indians of British Columbia used it as a treatment for epilepsy.

The Calypso Orchid relies on "pollination by deception", as it attracts insects which it does not nourish and which eventually begin to learn not to revisit it. Avoiding such recognition may account for some of the small variation in the flower's appearance.

Ceiba

Ceiba is the name of a genus of many species of large trees found in tropical areas, including Mexico, Central and South America, The Bahamas, the Caribbean, West Africa, and Southeast Asia. Some species can grow to 70 meters tall or more, with a straight, largely branchless trunk that culminates in a huge, spreading canopy, and "buttress" roots that can be taller than a grown man. The best-known, and most widely cultivated, species is Kapok, *Ceiba pentandra.* Recent botanical opinion incorporates *Chorisia* within *Ceiba,* raising to number of species from the previously accepted figure of 10 to as many as 20 or more, and puts the genus as a whole within the family Malvaceae.

Ceiba species are used as food plants by the larvae of some Lepidoptera species including the leaf-miner *Bucculatrix ceibae* which feeds exclusively on the genus.

Cistus

Cistus is a genus of flowering plants in the rockrose family Cistaceae, containing about 20 species. They are perennial shrubs found on dry or rocky soils throughout the Mediterranean region, from Morocco and Portugal through to the Middle East, and also on the Canary Islands. The leaves are evergreen, opposite, simple, usually slightly rough-surfaced, 2-8 cm long; in a few species (notably *C. ladanifer*), the leaves are coated with a highly aromatic resin called labdanum. They have showy 5-petaled flowers ranging from white to purple and dark pink, in a few species with a conspicuous dark red spot at the base of each petal., and together with its many hybrids and cultivars is commonly encountered as a garden flower.

The common name rockrose is applied to the species, a name also shared by the related genera *Halimium, Helianthemum* and *Tuberaria,* all in the family Cistaceae.

They are thermophilous plants, which require open, sunny places. As with many other Cistaceae, the species of *Cistus* have the ability to form mycorrhizal associations with truffles (*Tuber*) and are thus able to thrive on poor sandy soils or rocks.

Cistus are the only host of Cytinus hypocistis, a small parasitic plant that lives on the roots and is noticeable only for a short period of time when in flower. The presence of the parasite does not seem to hurt the host population.

Cistus species are used as food plants by the larvae of some Lepidoptera species including Coleophora confluella and Coleophora helianthemella, the latter recorded on Cistus monspeliensis.

Cosmos

Cosmos is a genus of about 20-26 species of annual and perennial plants in the family Asteraceae, native to scrub and meadow areas in Mexico (where the bulk of the species occur), the southern United States (Arizona, Florida), Central America and northern South America south to Paraguay.

They are herbaceous perennial plants growing 0.3-2 m tall. The leaves are simple, pinnate, or bipinnate, and arranged in opposite pairs. The flowers are produced in a capitulum with a ring of broad ray florets and a center of disc florets; flower color is very variable between the different species.

Selected species

- Cosmos atrosanguineus
- Cosmos bipinnatus
- Cosmos caudatus
- Cosmos diversifolius
- Cosmos herzogii
- Cosmos parviflorus
- Cosmos peucedanifolius
- Cosmos scabiosoides
- Cosmos sulphureus

The genus includes several ornamental plants popular in gardens, including Cosmos bipinnatus (also known as "Mexican Aster"), C. atrosanguineus ("Chocolate Cosmos") and Cosmos sulphureus ("Yellow Cosmos"). Numerous hybrids and cultivars have been selected and name

Dahlia

Dahlia is a genus of bushy, summer and autumn flowering, tuberous perennial plants native to Mexico, where they are the national flower. The Aztecs gathered and cultivated the dahlia for food, ceremony as well as decorative purposes, and the long woody stem of one variety was used for small pipes.

In 1872 a box of Dahlia roots were sent from Mexico to the Netherlands. Only one plant survived the trip, but produced spectacular red flowers with pointed petals. Nurserymen in Europe bred from this plant, which was named *Dahlia juarezii* with parents of Dahlias discovered earlier and these are the progenitors of all modern Dahlia hybrids. Ever since, plant breeders have been actively breeding Dahlias to produce thousands of cultivars, usually chosen for their stunning and brightly coloured flowers. Dahlia plants range in height from as low as 12" (30cm) to as tall as 6-8 feet (180-240cm). The flowers can be as small as 2" (5 cm) or up to a foot (30 cm) in diameter. The great variety results from Dahlias being octoploids (they have eight sets of homologous chromosomes, whereas most plants have only two).

Dahlias are used as food plants by the larvae of some Lepidoptera species including Angle Shades, Common Swift, Ghost Moth and Large Yellow Underwing. The dahlia is named after Swedish 18th-century botanist Anders Dahl.

Dendrobium

Dendrobium Swartz is a large genus of tropical orchids that consists of about 1200 species. The genus occurs in diverse habitats throughout much of south, east and southeast Asia, including the Philippines, Borneo, Australia, New Guinea and New Zealand. The name is from the Greek *dendron* (meaning tree) and *bios* (meaning life).

The species are either epiphytic, growing on a tree, or occasionally lithophytic, growing over a rock. They have adapted to a wide variety of habitats, from the high altitudes in the Himalayan mountains to lowland tropical forests and even to the dry climate of the Australian desert.

The orchids in this genus often develop pseudobulbs, which unite into a long reedlike stem with a typical length of more than 30 cm. Some appear densely covered with short white hairs. The short, ovate leaves grow alternately over the whole length of the stems. The axillary flower buds develop into short flower stalks with one or two terminal flowers. The orchids grow quickly throughout summer, but take a long rest during winter. In the spring, new shoots are formed from the base of the main plant and the dormant buds come back into action. The blooming flowers are found on pseudobulbs formed in the previous year.

Some species are in great demand by orchid lovers. This has resulted in numerous varieties and hybrids, such as the *Dendrobium nobile* varieties, which have greatly extended the range of colors of the original flower from the Himalayas. Kimilsungia, one of the national flowers of North Korea, is the cultivar *Dendrobium* 'Kim il Sung'.

Gymea Lily

The Gymea Lily (*Doryanthes excelsa*) is a flowering plant indigenous to the coastal areas of New South Wales near Sydney.

The plant has sword-like leaves more than a meter long. It flowers in spring and summer, sending up a flower spike up to 6 m high, which at its apex bears a large cluster of bright red flowers, each 10 cm across.

The name "Gymea Lily" is derived from a local Eora dialect. *Dory-anthes* means spear-flower in Greek, and *excelsa* is Latin for exceptional. The Sydney suburbs of Gymea and Gymea Bay are named after the lily. Its discovery in the Sydney basin region was recorded on 11th June 1853.

Honey-eaters love the nectar of these large, crimson flowers on stems 2-3 m tall. These birds were ready prey to be cooked and eaten! Aboriginal people in the Lake Macquarie district of NSW were observed in 1836 roasting the stems, having cut them when half a meter high and as thick as a person's arm. They also roasted the roots which they made into a sort of cake to be eaten cold.

Oil Palms

The oil palms (*Elaeis*) comprise two species of the Arecaceae, or palm family. They are used in commercial agriculture in the production of palm oil. The African Oil Palm *Elaeis guineensis* is native to west Africa, occurring between Angola and Gambia, while the American Oil Palm *Elaeis oleifera* is native to tropical Central America and South America.

Mature trees are single-stemmed, and grow to 20 m tall. The leaves are pinnate, and reach between 3-5 m long. A young tree produces about 30 leaves a year. Established trees over 10 years produce about 20 leaves a year. The flowers are produced in dense clusters; each individual flower is small, with three sepals and three petals. The fruit takes five to six months to mature from pollination to maturity; it comprises an oily, fleshy outer layer (the pericarp), with a single seed (kernel), also rich in oil. Unlike its relative, the Coconut Palm, the oil palm does not produce offshoots; propagation is by sowing the seeds.

Oil palms are grown for their clusters of fruit, which can weigh 40-50 kg. Upon harvest, the drupe, pericarp and seeds are used for production of soap and edible vegetable oil; different grades of oil quality are obtained from the pericarp and the kernel, with the pericarp oil used mainly for cooking oil, and the kernel oil used in processed foods.

For each hectare of oil palm, which is harvested year-round, the annual production averages 10 tonnes of fruit, which yields 3,000 kg of pericarp oil, and 750 kg of seed kernels, which yield 250 kg of high quality palm kernel oil as well as 500 kg of kernel meal. The meal is used to feed livestock. Some varieties have even higher productivities which has led to their consideration for producing the vegetable oil needed for biodiesel.

The African Oil Palm was introduced to Sumatra and the Malaya area in the early 1900s; many of the largest plantations of oil palms are now in this area, with Malaysia growing over 20,000 square kilometres. Malaysia claims that in 1995 it was the world's largest producer with 51% of world production. In this area, the destruction of natural rainforest to grow oil palm plantations is an issue of major environmental concern.

The oil palm is a tropical palm tree. There are two species of oil palm, the better known one is the one originating from Guinea, Africa and was first illustrated by Nicholaas Jacquin in 1763, hence its name, *Elaeis guineensis* Jacq.

The fruit is reddish about the size of a large plum and grows in large bunches. A bunch of fruits can weigh between 10 to 40 kilograms each. Each fruit contains a single seed (the palm kernel) surrounded by a soft oily pulp. Oil is extracted from both the pulp of the fruit (palm oil, an edible oil) and the kernel (palm kernel oil, used mainly for soap manufacture). For every 100 kilograms of fruit bunches, typically 22 kilograms of palm oil and 1.6 kilograms of palm kernel oil can be extracted.

Ficus

Ficus is a genus of about 800 species of woody trees, shrubs and vines in the family Moraceae, native throughout the tropics with a few species extending into the warm temperate zone.

The genus includes one species, the Common Fig *F. carica,* that produces a commercial fruit called a *fig;* the fruit of many other species are edible though not widely consumed. Other examples of figs include the banyans and the Sacred Fig (Peepul or Bo) tree. Most species are evergreen, while those from temperate areas, and areas with a long dry season, are deciduous.

A fig "fruit" is derived from a specially adapted type of inflorescence (structural arrangement of flowers). What is commonly called the "fruit" of a fig is actually a specialized structure- or accessory fruit- called a syconium: an involuted (nearly closed) receptacle with many small flowers arranged on the inner surface. Thus the actual flowers of the fig are unseen unless the fig is cut open. The syconium often has a bulbous shape with a small opening (the ostiole) at the distal end that allows access by pollinators. The flowers are pollinated by very small wasps that crawl through the opening in search of a suitable place to reproduce (lay eggs). Without this pollinator service fig trees cannot reproduce by seed. In turn, the flowers provide a safe haven and nourishment for the next generation of wasps.

The genus Dorstenia, also in the fig family (Moraceae), exhibits similar tiny flowers arranged on a receptacle but in this case the receptacle is a more or less flat, open surface.

Most figs come in two sexes: hermaphrodite (called caprifigs from goats - Caprinae subfamily; as in fit for eating by goats; sometimes called "inedible") and female. Fig wasps grow in caprifigs but not in the other because the female trees' female flower part is too long for the wasp to successfully lay her eggs in them. Nonetheless, the wasp pollinates the flower with pollen from the fig it grew up in, so figs with developed seeds also contain dead fig wasps almost too tiny to see.

When a caprifig ripens, another caprifig must be ready to be pollinated. Tropical figs bear continuously, enabling fruit-eating animals to survive the time between mast years. In temperate climes, wasps hibernate in figs, and there are distinct crops. Caprifigs have three crops per year; edible figs have two. The first of the two is small and is called breba; the breba figs are olynths. Some selections of edible figs do not require pollination at all, and will produce a crop of figs (albeit without fertile seeds) in the absence of caprifigs or fig wasps.

There is typically only one species of wasp capable of fertilizing the flowers of each species of fig, and therefore plantings of fig species outside of their native range results in effectively sterile individuals. For example, in Hawaii, some 60 species of figs have been introduced, but only four of the wasps that fertilize them have been introduced, so only four species of figs produce viable seeds there.

Figs are also easily propagated from cuttings. An extraordinarily large self-rooted Wild Willowleaf Fig in South Africa is protected by the Wonderboom Nature Reserve.

Strawberry

The strawberry (Fragaria) is a genus of plants in the family Rosaceae, and the fruit of these plants. There are more than 20 named species and many hybrids and cultivars. The most common strawberries grown commercially are cultivars of the Garden strawberry, Fragaria × ananassa. Strawberries are a valuable source of vitamin C. See Garden Strawberry for information about the fruit as a food.

The strawberry is an accessory fruit; that is, the fleshy part is derived not from the ovaries which are the "seeds" (actually achenes) but from the peg at the bottom of the hypanthium that held the ovaries. So from a technical standpoint, the seeds are the actual fruits of the plant, and the flesh of the strawberry is modified receptacle tissue. It is Whitish-Green as it develops and in most species turns red when ripe.

There are more than 20 different Fragaria species worldwide. Key to the classification of strawberry species is recognizing that they vary in the number of chromosomes. There are seven basic types of chromosomes that they all have in common. However, they exhibit different polyploidy. Some species are diploid, having two sets of the seven chromosomes (14 chromosomes total). Others are tetraploid (four sets, 28 chromosomes total), hexaploid (six sets, 42 chromosomes total), octoploid (eight sets, 56 chromosomes total), or decaploid (ten sets, 70 chromosomes total).

As a rough rule (with exceptions), strawberry species with more chromosomes tend to be more robust and produce larger plants with larger berries (Darrow).

Gardenia

Gardenia is a genus of about 250 species of flowering plants in the family Rubiaceae, native to the tropical and subtropical regions of Africa, southern Asia, Australasia and Oceania. The genus was named by Carolus Linnaeus after Dr Alexander Garden (1730-1791), Scottish-born American naturalist.

They are evergreen shrubs and small trees growing to 1-15 m tall. The leaves are opposite or in whorls of three or four, 5-50 cm long and 3-25 cm broad, dark green and glossy with a leathery texture. The flowers are solitary or in small clusters, white or pale yellow, with a tubular-based corolla with 5-12 lobes ('petals') from 5-12 cm diameter. Flowering is from about mid-spring to mid-summer and many species are strongly scented.

Gardenia plants are prized for the strong sweet scent of their flowers, which can be very large in some species.

Gardenias have a reputation for being difficult to grow. Because they originated in warm humid tropical areas, they demand high humidity to thrive. They flourish in acidic soils with good drainage. Potting soils developed especially for gardenias are available.

In Japan and China, Gardenia augusta is called Kuchinashi (Japanese) and Zhi zi (Chinese); the bloom is used as a yellow dye, which is used for clothes and food (including the Korean mung bean jelly called hwangpomuk).

Gordonia

Gordonia is a genus of flowering plants in the family Theaceae, related to Franklinia, Camellia and Stuartia. Of the roughly 40 species, all but one are native to southeast Asia in southern China, Taiwan and Indochina; the remaining species, G. lasianthus (Loblolly Bay), is native to southeast North America, from Virginia south to Florida and west to Louisiana.

They are evergreen trees, growing to 10-20 m tall. The bark is thick and deeply fissured. The leaves are alternately arranged, simple, serrated, thick, leathery, glossy, and 6-18 cm long. The flowers are large and conspicuous, 4-15 cm diameter, with 5 (occasionally 6-8) white petals; flowering is in late winter or early spring. The fruit is a dry five-valved capsule, with 1-4 seeds in each section.

The species are adapted to acidic soils, and do not grow well on chalk or other calcium-rich soils. They also have a high rainfall requirement and will not tolerate drought.

Some botanists include Franklinia within Gordonia; it differs in being deciduous and flowering in late summer, not late winter. The draft Flora of China account of Theaceae in China splits Gordonia into two genera, with G. lasianthus retained in Gordonia, and the Asian species transferred to Polyspora; this treatment is not yet widely accepted.

Several species of Gordonia are grown as ornamental plants for their flowers produced in winter when few other trees are in flower. They are however difficult to grow compared to the similar but generally smaller-growing camellias.

Helianthus (Sunlower)

The Helianthus L. genus comprises 67 species and several subspecies in the Asteraceae family, all of which are native to North America, with some species (particularly Helianthus annuus (sunflower) and Helianthus tuberosus (Jerusalem artichoke)) cultivated in Europe and other parts of the world as food crops and ornamental plants.

These are usually tall, annual or perennial plants, growing to a height of 60-390 cm. The rough and hairy stem is branched in the upper part. The petiolate leaves are dentate and often sticky. The lower leaves are opposite, ovate or often heart-shaped. The upper leaves are alternate and narrower.

They bear several to many wide, terminal capitula (flower heads), with bright yellow ray florets at the outside and maroon disc florets inside. These flower heads follow the direction of the sun, going from east to west during the day.

Jasmine

Jasmine or Jessamine (Jasminum) is a genus of shrubs and vines in the Family Oleaceae, with about 200 species, native to tropical and warm temperate regions of the Old World. The majority of species grow as climbers on other plants or on structures. The leaves can be either evergreen or deciduous, and are opposite in most species; leaf shape is simple, trifoliate or pinnate with up to nine leaflets.

Jasmine flowers are generally white, although some species have yellow flowers. Unlike most genera in the Oleaceae which have four corolla lobes ("petals"), jasmines often have five or six lobes. They are often strongly and sweetly scented. Flowering is in spring or summer in most species, but in a few species, notably J. nudiflorum, in winter on the bare branches of this deciduous species.

The common name "jasmine" is often given to unrelated plants with pale, sweetly-scented flowers and dark green leaves, such as Trachelospermum species (Confederate or star jasmine), Gardenia jasminoides (Cape jasmine), Cestrum nocturnum (Night blooming jasmine) and Gelsemium species (Carolina jasmine).

Jasmines are widely cultivated for their flowers, enjoyed in the garden, as house plants, and as cut flowers. The flowers are worn by women in their hair in southern and southeast Asia. Some claim that the daily consumption of Jasmine tea is effective in preventing certain cancers. Many species also yield an absolute which is used in the production of perfumes and incense.

Jasminum sambac flowers are also used to make tea, which typically has a base of green tea, but sometimes an Oolong base is used. The delicate Jasmine flower opens only at night and is plucked in the morning when the tiny petals are tightly closed.

They are then stored in a cool place until night. Between six and eight in the evening, as the temperature cools, the petals begin to open. Flowers and tea are "mated" in machines that control temperature and humidity. It takes four hours or so for the tea to absorb the fragrance and flavour of the Jasmine blossoms, and for the highest grades, this process may be repeated as many as seven times. Because the tea has absorbed moisture from the flowers, it must be refired to prevent spoilage. The spent flowers may or may not be removed from the final product, as the flowers are completely dry and contain no aroma. They simply add visual appeal and are no indication of the quality of the tea.

Jasminum sambac is also the National Flower of Indonesia, where it is known as "Melati", and of the Philippines, where it is known as "Sampaguita". In Indonesia (especially the island of Java), it is the most important flower in wedding ceremonies for ethnic Indonesians. Jasminum officinale is the national flower of Pakistan, where it is known as the "Chameli" or "Yasmine". In Sanskrit it is called Mallika. Jasmine is cultivated at Pangala, in Karnataka, India, and exported to Middle East countries.

Jatropha

Jatropha is a genus of approximately 175 succulents, shrubs and trees (some are deciduous, like Jatropha curcas L.), from the family Euphorbiaceae. Plants from the genus natively occur in India, Africa, North America, and the Caribbean.

Originating in the Caribbean, the jatropha was spread as a valuable hedge plant to Africa and Asia by Portuguese traders. The mature small trees bear male and female inflorescence, and do not grow very tall.

Jatropha curcas seeds is widely used for making biodiesel fuel in India, and is being promoted as an easily grown biofuel crop in hundreds of projects throughout India and the third world. The rail line between Mumbai and Delhi is planted with Jatropha and the train itself runs on 15-20% biodiesel.

The plant can grow in wastelands, and it yields more than four times as much fuel per hectare as soybean, and more than ten times that of corn. A hectare of jatropha produces 1,892 liters of fuel. Also Jatropha is a one-stage conversion to biodiesel.

Kohleria

Kohleria is a New World genus of the flowering plant family Gesneriaceae. The plants are generally tropical herbs or subshrubs with velvety stems and foliage and brightly colored flowers with spots or markings in contrasting colors. They are rhizomatous and commonly include a period of dormancy in their growth cycle. The genus was recognized as having 19 species distributed in Central America and South America. Recent phylogenetic work by Roalson et al. (2005) indicated that the epiphytic genus Capanea is derived from within Kohleria, and the two species of Capanea were subsequently transferred to Kohleria. The genus Pearcea is closely related.

Because of their colorful and exotically patterned flowers, as well as a general interest in the many tropical flowering plants that were being introduced for the first time from the Americas, kohlerias were very popular in England and Europe in the 19th Century. Many species and hybrids were lavishly illustrated in horticultural magazines such as Curtis's Botanical Magazine under the discarded or erroneous names of *Achimenes, Gesneria, Isoloma, Sciadocalyx,* and *Tydaea.* These species and hybrids almost entirely disappeared in the early 20th Century, and plant breeders have only recently begun to work extensively with this genus again.

Several species are widespread, variable, weedy, and tend to hybridize in the wild, and numerous names have been described that are synonyms of other species or are hybrid taxa.

Lilium(Lily)

The plants of the genus Lilium are the true lily plants, comprising a genus of about 100 species in the lily family, Liliaceae. They are important as showy and large flowered garden plants, and in literature.

Lilies are native to the northern temperate regions. Their range in the Old World extends across much of Europe, the north Mediterranean, across most of Asia to Japan, south to the Nilgiri mountains in India, and south to the Philippines. In the New World they extend from southern Canada through much of the United States. A few species formerly included within this genus have now been placed in other genera. These include Cardiocrinum and Nomocharis.

They are commonly adapted to either woodland habitats, often montane, or sometimes to grassland habitats. A few can survive in marshland and a single one is known to live as an epiphyte (L. arboricola). In general they prefer moderately acidic or lime-free soils.

Lilies are usually erect leafy stemmed herbs. The majority of species form naked or tunic-less scaly underground bulbs from which they overwinter. In some North American species the base of the bulb develops into rhizomes, on which numerous small bulbs are found. Some species develop stolons. A few species form bulbs at or near the soil surface .

Many species form stem-roots. With these, the bulb grows naturally at some depth in the soil, and each year the new stem puts out adventitious roots above the bulb as it emerges from the soil. These roots are in addition to the basal roots that develop at the base of the bulb.

The large flowers have three petals along with three petal-like sepals, often fragrant, and come in a range of colours ranging through whites, yellows, oranges, pinks, reds, purples, bronze and even nearly black. Markings include spots, brush strokes and picotees.

The plants are summer flowering. Most species are deciduous, but Lilium candidum bears a basal rosette of leaves for much of the year. Flowers are formed at the top of a single erect stem, with leaves being borne at intervals up the stem.

Mentha (Mint)

Mentha (mint) is a genus of about 25-30 species of flowering plants in the family Lamiaceae, with a subcosmopolitan distribution, seven from Australia, one in North America, and the others from Europe and Asia; several hybrids also occur.

They are aromatic perennial herbs, growing to 10-120 cm tall, with wide-spreading underground rhizomes and erect, branched stems. The leaves are arranged in opposite pairs, simple oblong to lanceolate, often downy, and with a serrated margin. The flowers are produced in clusters ('verticils') on an erect spike, white to purple, the corolla two-lipped with four subequal lobes, the upper lobe usually the largest. The fruit is a small dry capsule containing 1-4 seeds.

Mints are growing vigorous, spreading plants that tolerate a wide range of conditions, but thrive where there's abundance of water. They can be highly invasive plants, so caution should be taken in cultivation or it can take over an entire garden.

The most common and popular mints for cultivation are peppermint (Mentha × piperita), spearmint (Mentha spicata), and (more recently) pineapple mint (Mentha suaveolens). The dark green leaves have a pleasant warm, fresh, aromatic, sweet flavor with a cool aftertaste. Mint essential oils are used to flavor food, candy, teas, breath fresheners, antiseptic mouth rinses, and toothpaste. Mint leaves are used in teas, beverages, jellies, syrups, and ice creams. In Mid-Eastern cuisines, mint is used on lamb dishes. In British cuisine, mint sauce is popular with lamb. Mint leaves generally refer to spearmint leaves. Pennyroyal and Corsican Mint have a much stronger odor and flavor, and potentially harmful medicinal effects. Pennyroyal resembles the other mints, but Corsican mint is unusual in the fact that it is a low, mossy ground-covering plant. The pineapple mint is particularly mild and popular.

Mint essential oil and menthol are extensively used as flavorings in drinks, chewing gum and desserts/candies; see mint (candy) and mint chocolate. The substances that give the mints their characteristic aromas and flavors are:

- Menthol: the main aroma of Spearmint, Peppermint, and Japanese Peppermint (a major commercial source).
- Pulegone: in Pennyroyal and Corsican Mint.

Nelumbo(Lotus)

Nelumbo is a genus of water flowers commonly known as lotus and the only genus in the family Nelumbonaceae. Nelumbo is what is most commonly meant by the rather ambiguous term "lotus". The word Nelumbo originates from the Sinhalese word Nelum.

Nelumbonaceae is the botanical name for a family of flowering plants. Such a family has been recognized by only a few taxonomists: the plants involved have often been included in family Nymphaeaceae. It is sometimes called the "lotus family" or "lotus-lily family".

Nelumbo is superficially similar to but not closely related to the water-lilies of the family Nymphaeaceae. The leaves of Nelumbo can be distinguished from those of genera in the Nymphaeaceae as they are peltate, that is they have fully circular leaves. Nymphaea, on the other hand, has a single characteristic notch from the edge in to the center of the lilypad. The central seed pod of Nelumbo is also a distinguishing feature.

Nelumbo nucifera is particularly well known as it is the sacred lotus of Hinduism and Buddhism and is the national flower of India. Its roots are also used widely in Asian cooking.

Osmunda

Osmunda is a genus of primarily temperate-zone ferns of family Osmundaceae. Five to ten species have been listed for this genus.

The species have completely dimorphic fronds or pinnae, green photosynthetic sterile fronds, and non-photosynthetic spore-bearing fertile fronds, with large, naked sporangia. Because of the large mass of sporangia that ripen uniformly at the same time to a showy golden color, the ferns look as if they are in flower, and so this genus is sometimes called the "flowering ferns".

Selected species

- Osmunda asiatica (syn. Osmunda cinnamomea var. asiatica) – Asian Cinnamon Fern
- Osmunda bromeliifolia
- Osmunda cinnamomea – Cinnamon Fern
- Osmunda claytoniana – Interrupted Fern, Clayton's Fern
- Osmunda japonica – Japanese Flowering Fern
- Osmunda regalis – Royal Fern

Osmunda species are used as food plants by the larvae of some Lepidoptera species including The Engrailed.

One of the species, the Cinnamon fern (Osmunda cinnamomea) forms huge clonal colonies in swamp areas. These ferns form massive rootstocks with densely-matted, wiry roots. This root mass is an excellent substrate for many epiphytal plants. They are often harvested as osmunda fiber and used horticulturally, especially in propagating and growing orchids. The genus name is from Middle English and Middle French words for a type of fern.

Oak

The term oak can be used as part of the common name of any of several hundred species of trees and shrubs in the genus Quercus (from Latin "oak tree"), and some related to jessica, notably Cyclobalanopsis and Lithocarpus. The genus is native to the northern hemisphere, and includes deciduous and evergreen species extending from cold latitudes to tropical Asia and the Americas.

Oaks have spirally arranged leaves, with a lobed margin in many species; some have serrated leaves or entire leaves with a smooth margin. The flowers are catkins, produced in spring. The fruit is a nut called an acorn, borne in a cup-like structure known as a cupule; each acorn contains one seed (rarely two or three) and takes 6-18 months to mature, depending on species.

Willows

The willows are deciduous trees and shrubs in the genus Salix, part of the willow family Salicaceae. There are about 350 species in this genus worldwide, found primarily on moist soils in cooler zones in the Northern

Hemisphere. The leaves are deciduous, often elongate but round to oval in a few species, and with a serrated margin. Willows are dioecious with male and female flowers appearing as catkins on different plants; the catkins are produced early in the spring, often before the leaves or as the new leaves open. The fruit is a small capsule containing numerous tiny (0.1 mm) seeds embedded in white down, which assists wind dispersal of the seeds. Willows are very cross-fertile and numerous hybrids are known, both naturally occurring and in cultivation.

Some smaller species may also be known by the common names osier and sallow; the latter name is derived from the same root as the Latin salix.

Some willows, particularly arctic and alpine species, are very small; the Dwarf Willow (Salix herbacea) rarely exceeds 6 cm in height, though spreading widely across the ground.

The Weeping Willow, very widely planted as an ornamental tree, is a cultivar, Salix × sepulcralis 'Chrysocoma', derived from a hybrid between the Chinese Peking Willow and the European White Willow.

Almost all willows take root very readily from cuttings or where broken branches lie on the ground. There are a few exceptions, including the Goat Willow and Peachleaf Willow. One famous example of such growth from cuttings involves the poet Alexander Pope, who begged a twig from a parcel tied with twigs sent from Spain to Lady Suffolk. This twig was planted and thrived, and legend has it that all of England's Weeping Willows are descended from this first one. Willows are used as food plants by the larvae of some Lepidoptera species. A number of willow species were widely planted in Australia in the past, notably as erosion control measures along watercourses.

Sobralia

Sobralia is a genus of about 125 orchids (family Orchidaceae) and the only genus of the subtribe Sobraliinae. It is native to Central and South America. The plants are more commonly terrestrial, but are also found growing epiphytically, in wet forests from sea level to about 8,800 ft. The genus was named for Dr. Francisco Sobral, a Spanish botanist.

Their reed-like stems range in height from about 1 ft (such as in Sobralia galeottiana) to 25-30 ft. They have typically heavily veined, bilobed, plicate, apical leaves all along the stem. The inflorescenses on the apex of the stem carry one or two successive ephemeral flowers with large sepals and petals. The short duration of the flower is caused by a self-digesting enzyme. The lip is entire or lobed and clasps the column at its base. This columns carries eight soft pollinia. These flowers range in color from pure white to yellow, green, pink, purple, red, brown, and even a blue violet.

Umbellularia Californica

Umbellularia californica is a tree native to coastal forests of western North America. Its pungent leaves have a similar flavor to bay leaves (though stronger), and it may be mistaken for Bay Laurel.

In Oregon, this tree is known as Oregon Myrtle, while in California it is called California Bay Laurel, which may be shortened to California Bay or California Laurel. It has also been called Pepperwood and Headache Tree (the latter from the strong scent of the crushed foliage, which can cause a headache if breathed in to excess).

It is the sole species in the genus Umbellularia. It ranges near the coast from Douglas County, Oregon south through California to San Diego County. It is also found in the western foothills of the Sierra Nevada mountains. It occurs at altitudes from sea level up to 1600 m.

It is an evergreen tree growing to 30 m tall (exceptionally 45 m) with a trunk up to 80 cm thick. The leaves are smooth-edged and oval, 3–10 cm long and 1.5–3 cm broad, similar to the related Bay Laurel though usually narrower, and without the crinkled margin of that species. The flowers are small, yellow or yellowish-green, produced in a small umbel (hence the scientific name Umbellularia, "little umbel"). The fruit is a round and green berry 2–2.5 cm long and 2 cm broad, lightly spotted with yellow, maturing purple. It consists of a fleshy covering over a single hard, thin-shelled seed.

Zygopetalum

Zygopetalum is a genus of the orchid family (Orchidaceae) (subfamily Epidendroideae, tribe Maxillarieae, subtribe Zygopetalinae), consisting of fourteen species.

This orchid's generic name, derived from the Greek word "zygon", means "yoked petal." It refers to the yoke-like growth at the base of the lip caused by the fusion of petals and the sepals. They occur in humid forests at low- to mid-elevation regions of South America, with most species in Brazil. Most are epiphytes, but some are terrestrials with glossy, strap-like, plicate leaves, which are apical, oblong or elliptic-lanceolate, acute or acuminate. These orchids have a robust growth form. Their ovoid-conical pseudobulbs are deciduous.

They produce an erect, 60 centimeter-long, few-flowered to several-flowered, racemose inflorescence that grows laterally and is longer than the leaves. Their prominent bracts equal the length of the ovary. They are known for their fragrant, waxy, and long-lived flowers with multiple blooms in shades of green, purple, burgundy, and raspberry with several patterns. They are known for their ease of culture and are much in demand as excellent cut flowers.

INTEGRATED DISEASE MANAGEMENT

Effective, practical disease management usually involves several techniques. Management programs based on only one or two techniques can be effective in the short term but may become ineffective if used frequently. The development of fungicide resistance is a good example; the epidemics of rust caused by newly prevalent races of the fungus is another. Growers need to integrate as many different management tools as possible for long term success.

Stem rust of wheat is managed by host resistance, the elimination of barberry (this prevents the sexual phase that produces new rust races), and by constant monitoring of the rust races by plant pathologists. Monitoring provides early warning for plant breeders of the buildup of new rust races and the need for alternative management for the grower.

The main tools for management of dry bean rust are sanitation, crop rotation, tolerant varieties and fungicides. All of these procedures must be done well and in a timely fashion to produce effective management.

Tan spot of wheat can be reduced by sanitation, crop rotation, and protection (fungicides); researchers are developing varieties with improved tan spot resistance. When wheat is planted directly into standing wheat stubble, both sanitation and rotation are eliminated as management tools and growers must rely primarily on variety choice and fungicides for tan spot reduction.

Homeowners should make every effort to remove diseased plant debris from the yard and garden in the fall-sanitation is the first step to disease management for next spring. Sanitation should be followed by seed treatment, rotation, use of resistant varieties, and timely fungicide applications. Whenever possible, all the principles of management, eradication, exclusion, host resistance, and protection, should be practiced. The use of these combined practices usually produces the most reliable and stable plant disease management.

Finally, all growers should keep close watch on the weather forecasts, as this helps to anticipate disease-favouring weather and allows more timely preventive measures. Some Disease Forecasting models for major crop diseases have been developed and are now available on the Internet or through toll-free telephone numbers. The disease forecasting information aids in predicting disease outbreaks and the need for fungicides.

DISEASE CONTROL IN THE LANDSCAPE

Plants must be inspected regularly to detect any problems. When problems are discovered, appropriate actions must be taken to prevent or alleviate damage. Landscape managers have numerous management tools available; the key to their successful use is knowing when and how to apply them. The best approach is an integrated program that includes regular monitoring and relies on a combination of techniques to prevent and control problems.

Methods include selecting plants that are well adapted to the environment and resistant to plant diseases and pests. Our awareness of the importance of various cultural practices for preventing or controlling plant diseases as well as abiotic disorders has increased greatly in recent years. Especially in landscapes with an emphasis on reduced pesticide use, cultural practices are an extremely important and integral part of plant health management. Pesticides (primarily fungicides) may be an important component of disease management; however, this bulletin does not make specific recommendations regarding the use of fungicides because availability, appropriate, and legal uses of fungicides frequently change. General comments related to fungicide use are made for several diseases in the text; however, for specific recommendations, growers are referred to other sources of information (Useful References) that are more frequently updated than this bulletin, and to the ultimate legal source of information—The Pesticide Label.

GENERAL PRINCIPLES OF PLANT HEALTH MANAGEMENT

The basic requirements for growing healthy plants are relatively easy to identify. These include: good soil structure, texture and drainage; proper balance and amount of nutrients; acceptable soil pH; sufficient space for expanding roots and crowns; sufficient water, acceptable temperature and light; clean air; and relative freedom from pests and diseases. Significant fluctuations within or among any of the above mentioned environmental conditions can result in poor plant health.

Much of the information presented in this bulletin refers to the diagnosis and control of infectious plant diseases. However, it is important to remember that "sick" or unhealthy plants may not be infected with a plant pathogen.

In fact, many plant health problems encountered in Ohio landscapes are not related to plant pathogens, insects or other pests. Unfavourable environmental conditions in addition to many other factors can lead to plant stress. Depending upon the level and duration of stress, plant health can be greatly affected. If stress is severe and long term, it can lead to visible "abnormal" or "undesirable" symptoms and, eventually, death of the plant. Understanding the importance of plant stress, combined with the ability to identify stressful situations in order to prevent or alleviate them, is essential to successful plant health management in the landscape.

Although we emphasize the importance of stress on plant health, it is important to remember that a stress-free, perfectly healthy plant can be attacked by many plant pathogens. If a healthy plant is highly susceptible and environmental conditions are favourable for disease development, the pathogen can infect and cause disease that damages or even kills the plant. Although most plant pathogens do not require a stressed or weakened plant in order to infect, stressed or weakened plants may be more susceptible to attack by some plant pathogens. Some plant pathogens do require some degree of stress or plant injury before they can infect and cause disease. Examples of this are species of the fungus *Botryosphaeria*. These fungi can infect a wide range of woody ornamentals, but are rarely a problem on vigorous, healthy plants. On crabapple, it generally becomes a problem (stem cankers) after the tree has been predisposed to infection by some sort of injury. A common form of predisposing injury in Ohio is cold weather or winter injury. This type of relationship is well documented for several plant diseases. Thus, an understanding of the factors that predispose plants to disease is critical to successful plant health management.

CHRONIC AND ACUTE PLANT STRESS

Compacted, poorly-drained soils, nutrient imbalances, unsuitable soil pH, air pollution, limited air circulation/plant crowding, road salt injury, improper plant installation practices, inappropriate light levels, and poor irrigation practices are among the most common factors that create plant stress and may result in increased susceptibility to certain diseases. These factors are chronic stresses that weaken the plant over time. Months, years, or decades may be involved before the cumulative effects result in visible problems. Sometimes the chronic stress itself over a period of time will cause the plant to decline and eventually die without involvement of a pathogenic organism. These types of problems that are not caused by pathogenic organisms are generally referred to as 'disorders,' or 'abiotic' or 'noninfectious' diseases.

The prevention of infectious as well as non-infectious diseases linked to chronic stress from environmental factors depends on wise landscape planning and management. Choosing appropriate plants and materials for the site, appropriate site preparation and careful maintenance will go a long way towards avoiding plant health problems. An understanding of the cultural requirements of landscape ornamentals is essential to alleviating plant health problems related to chronic stress.

Acute environmental stress is linked to a single event and, like chronic stress, may sometimes predispose a plant to disease. Damage from drought, cold, wind, hail, and mechanical damage are all examples of acute stress. Unlike chronic stress that occurs over time, acute stress occurs suddenly and causes damage soon after it occurs. With chronic stress, there may be time to reverse an imbalance if it is recognized early; however, with acute stress there is little or no time to prevent plant damage.

Diagnosing Plant Health Problems

Problems of ornamentals usually do not arise from an isolated cause. There may be a primary cause such as spider mites on spruce or rust disease on hawthorn. However, associated environmental conditions may be present that need to be diagnosed. Drought (for instance) often precedes mite problems. Formulating questions to ask about the ornamental plant with poor health symptoms is a key to successful diagnosis.

One of the most difficult diagnostic areas in ornamental plant care is determining the causes of problems that result in nonspecific symptoms. Such nonspecific symptoms as leaf yellowing, leaf drop, browning of the tips of leaves and branch diebacks may be serious and easily detected. However, they can result from a variety of environmental imbalances, alone or in combination, as well as from some kinds of pathogens.

Such nonspecific symptoms commonly, but not always, result from poor root health in the urban landscape. Hundreds of factors can cause root health problems. Some of these may be infectious diseases and others may be noninfectious environmental problems. The only way to achieve a correct determination of the causes of a plant problem when beginning with a nonspecific symptom is to try and find more symptoms and information. Gather more data in order to produce a symptom profile. Even if the set of symptoms includes nothing more than a group of individually nonspecific conditions, it might still lead to a correct diagnosis.

Suppose you start with yellow leaves. You could increase the symptom profile by: 1) noting where on the plant the yellow leaves are, 2) when they first appeared, 3) what unusual cultural practice might have occurred, 4) what other leaf conditions you see (browning, leaf drop, leaf size change), 5) what soil conditions are present (wet, dry, compacted, acidic, salty). Perhaps the help of an outside diagnostic clinic or a soil testing service will be needed. The result will be a greater ability to correctly diagnose the cause of the symptom and plan a solution.

There is no substitute for thorough field observation. When confronted with a plant health problem, be prepared to look at the plant and the surrounding environment closely. A good way to proceed is to answer a series of questions. Remember that each question can have one or more answers. The following paragraph outlines an approach that has worked for many.

First, properly identify the plant. With this knowledge you can determine the cultural requirements for the plant. Knowing the plant type will also enable you to anticipate what specific diseases to expect and when to expect them.

For instance, if you have a dead branch on a tree, you might expect first fireblight if it's a crabapple or Verticillium wilt if it's a Norway maple. Gather reference material and be familiar with the symptom profile to be expected. Remember, pictures in texts often are out of perspective because they are selected for their ease of showing one problem. The plant you are observing may exhibit symptoms of several diseases, environmentally induced stresses and insect damage all at once.

Other questions to ask during a diagnosis include the following:

1. What is wrong? Exactly what symptoms are of concern?
2. What is the condition of the surrounding vegetation?
3. How long has the problem been there? Be careful! Many people believe plants become sick overnight. Generally, plant health problems develop over considerable time.
4. What are all the possible causes of the problem? Be sure to list them, including environmental conditions and anything else you think might be contributing to the situation. If necessary, have a plant disease clinic assist you.
5. What are the remedies or health management practices that can be considered? Next to the causes, list all possible remedies, regardless of cost or practicality.
6. Of the possible remedies, which are practical? This is a crucial step! Remember, doing nothing and merely tolerating the situation may be the "most practical" thing to do.

7. Of the practical remedies, when would be the best time of year to apply them? Will any follow-up activity such as reapplication of a spray be needed?

Using the Plant and Pest Diagnostic Clinic

The C. Wayne Ellett Plant and Pest Diagnostic Clinic (PPDC) is located at The Ohio State University in Columbus, Ohio. It is a comprehensive clinic in that it integrates the disciplines of plant pathology, entomology, crop science, horticulture, and natural resources into a comprehensive program centred on plant disease and pest identification. Services available through PPDC include plant disease diagnosis, insect and mite identification, plant and weed identification, and nematode extractions.

The accuracy of any diagnosis depends upon the information supplied, the specimen material selected, and the condition of the specimen when it arrives.

Specimen Information

PPDC specimen forms are available through county Extension agents or at the clinic. To make best use of the forms:

1. Complete a form for each specimen. Materials that are part of the same problem will be considered as one specimen. Different kinds of plants or plants from different sites usually have different problems and should be considered as different specimens.
2. Describe symptoms as you observe them on the plant or in the planting. After a specimen is packaged and shipped, many of the original symptoms may be obscured.
3. Send the completed form along with the specimen, either inside the package or in an envelope attached to the outside of the package. Make sure that the record form and other enclosures are separated from the specimen material by a waterproof barrier if needed.

Selecting Specimen Material

Select material showing the symptoms in question. Send several samples in different stages of disease development. Be generous, but do not send totally rotten or decayed material, plants that are completely dead, or material that has been dead a long time.

1. Send entire plants, including roots if possible. Dig plants to keep roots intact. Do not pull them out of the ground. Remove excess soil from the roots by shaking gently or washing with water. Do not wet leaves

or stems. Wrap roots so that soil clinging to roots will not be loose in the package. Do not ship wet plants.

2. When entire plants cannot be shipped, include the part showing symptoms. Also, include one pint of roots, soil, and fine rootlets.
3. When only localized parts of plants are affected (leaf spots, stem cankers, or swellings), ship several examples of the affected parts. Stem and branch sections should include a short section of healthy tissue so that the transition area between diseased and healthy tissue is included.
4. If oak wilt, Dutch elm disease, or Verticillium wilt are suspected, select samples from branches with wilted, yellow, or dying leaves. Do not select samples that are completely dead. Cut several sections 6 inches long and one-half to 1 inch in diameter.

Packaging and Delivery of Specimens

Careful packaging and quick delivery of the specimens is essential. Please follow these suggestions:

1. Select specimens fresh from the field. Carefully shake off dew or excess water and allow to air-dry before packing.
2. Press non-woody plants or leaves on small branches between paper and place between two pieces of stiff cardboard. Place the sample in a padded envelope.
3. Soft, succulent plants that are packed in airtight plastic frequently decay before their arrival. Package these plants so they may "breathe." Place leaves between dry paper towels before packing.
4. For mailing, use strong containers such as corrugated boxes or mailing tubes that will not crush in transit.
5. Fill empty spaces in mailing cartons with crushed or shredded paper to protect the specimen.
6. Use overnight mail services or mail packages early in the week to avoid weekend layovers at the post office.

Recognizing and Understanding Infectious Plant Diseases

Infectious plant diseases differ from noninfectious or abiotic disorders because they are caused by specific plant pathogens. Plant pathogens are usually microorganisms that are on or inside the plant (the host).

Many plant pathogens are common in the air or soil around the host plants. Although plant pathogens may be common in many landscape situations, they cannot infect the plant and cause disease unless the environment is favourable for the interaction to take place. Thus, in order to have an infectious plant disease, a susceptible host plant, a pathogen and a favourable environment for disease development are required. These three elements make up what is referred to as the disease triangle Understanding this basic principle about infectious plant disease is important to developing successful disease management programs.

Many of the cultural practices that are useful in disease management are directed toward changing or creating an environment that is less conducive to disease development. Another important difference between infectious and noninfectious disease is that infectious diseases can spread from plant to plant by means of pathogen dispersal. Noninfectious diseases do not spread in this manner.

TYPES OF PLANT PATHOGENS

The most common plant pathogens on ornamental plants in Ohio are fungi, bacteria, viruses, and nematodes. Fungi are by far the most common type of plant pathogens. Plant pathogenic fungi cause many different symptoms, including leaf spots, wilts, curled leaves, dieback, enlargements or galls, and stunted or dead plants. Some fungi are visible in forms—called signs—such as rusts, mildews, and sooty molds. Fungi are usually composed of fine threadlike structures (hyphae) that form a network or mass (mycelium) growing on or through their host. Fungi can spread through rhizomorphs, sclerotia, or by microscopic spores. Rhizomorphs are rootlike or cordlike masses of hyphae that can contaminate soil and plant parts. Sclerotia are compact masses of hyphae that can persist for relatively long periods in soil.

Fungal spores (which are produced in great numbers) can be spread by wind, water, soil movement, machinery, insects, or other things with which they come in contact, including people. Sclerotia, rhizomorphs, and some types of spores may survive for long periods in or on plants or in soil. When and where conditions such as temperature, moisture, and the presence of a host plant are suitable for growth, they produce new fungal hyphae and infect the plant. If they are large enough to be seen with the naked eye or a hand lens, signs such as mycelium, masses of spores, and spore-forming structures such as mushrooms, help in identifying fungi.

Bacteria are microscopic, one-celled organisms that obtain their food from living plants or other organic matter. Common symptoms of disease resulting from bacterial infection are shoot blight, leaf spots, soft rots, scabs, wilts, cankers, and galls on branches, twigs, stems, and roots. Unlike fungi, plant pathogenic bacteria generally do not produce spores that can survive adverse environmental conditions; they must usually remain in contact with a host plant or plant debris to survive. Plant-infecting bacteria generally require warmth and moisture to multiply, and are usually not a problem during dry summer weather, except where there is overhead irrigation. Bacteria are commonly spread by splashing water, but may also be dispersed by insects or by moving infested plants, soil, or equipment. In some instances, such as diseases of herbaceous annual flowers, bacteria may be seed-borne.

Viruses are submicroscopic particles that can infect plants and lead to stunting, discoloring, deformation, or death of leaves, stems, fruit, or entire plants. Viruses rarely kill woody plants, and some infected plants exhibit no symptoms. Viruses require a living host cell in which to reproduce, and generally do not survive for very long outside of living tissue. Many viruses are spread by aphids, leafhoppers, or other plant-feeding insects. Some viruses are spread by nematodes, budding or grafting, or the movement of infected seeds, plants, plant parts, or infested equipment. Once a plant is infected by a virus, it usually remains infected during its entire life.

Nematodes are tiny (usually microscopic) roundworms that feed on a wide diversity of organisms. Nematode species that damage plants may feed on or in roots, tubers, bulbs, leaves, or stems. With the exception of the pinewood nematode, which feeds inside branches and trunks, most species that are pests of ornamental plants feed on roots. These include root knot nematodes, cyst nematodes, root lesion nematodes, and dagger, pin, ring and stunt nematodes.

TYPES OF INFECTIOUS PLANT DISEASES OF ORNAMENTALS

Powdery Mildews

Powdery mildew diseases are caused by fungi that are often specific to the host they infect. Generally, each plant type is unique in that the powdery mildew fungus that infects it will not infect any other. Powdery mildew is common on roses, lilacs, English oak, sycamore, some deciduous azaleas and zinnias, and many other plants

The white growth on leaves and stems is the powdery mildew fungus growing on the surface of the plant tissue. Much of this growth consists of spores, which can be blown about to susceptible plants nearby. Small structures, called haustoria, grow within host cells, injuring them as they obtain food. Powdery mildew will not usually kill a plant but may weaken it and reduce winter hardiness. The unsightly fungal growth greatly reduces the aesthetic quality of the plant. If the infection is widespread, the entire leaf blade may be covered with a thin, white fungal growth. Some mildew fungi affect older leaves first, such as on lilacs. Others affect newer shoots, such as on roses or crabapples. When new shoots are affected, leaf curling and shoot stunting and twisting may be severe.

Powdery mildew can be treated effectively with chemicals. The white appearance may remain, however, even though the fungus may be dead. For highly susceptible plants such as roses, a regular preventive spray program should be planned and carried out. For other plants, watering only early in the day and pruning out surrounding vegetation to improve sunlight and drying of plant foliage may be all that is needed for control.

Rust Diseases

Rust diseases are also caused by fungi that are highly specific to the host plant they infect. Rust fungi usually sporulate on plant tissues forming masses of orange to dark red spores. Orange to reddish-brown spots or lesions will form on leaves, twigs, or fruit. Sometimes galling and disfigurement will be associated with the infections. In summer, infections on some plants will produce cup-like structures or tubular growths that will bear abundant orange spores. Commonly affected trees are hawthorns, crabapples, or junipers. Some kinds of rust fungi must have two different species (hosts) of plants in the area before they become a problem. An example is Cedar-Apple rust, where both a rosaceous host (crabapple, hawthorn) and junipers are required for disease development. Some other common rust diseases include rust on aster, hollyhock, and snapdragon.

Leaf Spotting and Blighting

Most of these diseases are also caused by fungi; however, bacteria can also be the cause. Common bacterial diseases include bacterial leaf spot of English ivy and zinnia; fireblight on crabapple, mountain ash, hawthorn, cotoneaster, and pyracantha; crown gall on rose and euonymus; and bacterial leaf scorch of shade trees. The fungi that cause leaf spotting and blight are common and are usually spread by microscopic spores either long distances by air or shorter distances by splashing water. Most notable among this group are Entomosporium leaf spot on hawthorn, scab on crabapple and pyracantha, Botrytis blights, black spot of rose, and anthracnose on shade trees.

Anthracnose diseases are common and often result in irregular tan or brown areas on the leaves, especially along major veins of the leaf blade. The condition usually is noticed in late spring on lower parts of the plant. Deformation and twisting of leaves often results from infections that occurred during bud break and leaf expansion. Anthracnose blight is commonly found on ash, dogwood, white oak, maple, sycamore, and London plane tree and others, particularly when such plants are growing in damp, cool places. Frost injury, herbicide damage, and leaf scorch are often mistaken for anthracnose infection. Examine affected leaves closely for fungal fruiting bodies within or near the edge of browned areas and on the veins. Anthracnose is also common on walnut, causing early leaf fall later in the summer. On river birch, small circular leaf spots are the symptoms. On some hosts, especially sycamore and flowering dogwood, twig blight and cankers may occur.

Root and Crown Rots Caused by Water Molds

Pythium and *Phytophthora* fungi are often called water molds because they have a spore stage that is adapted to spread by swimming in water. These organisms attack a wide variety of plants, causing root rots, stem rots, and cutting rots. Many times they do not kill a plant. Instead, they may "prune" the root system, resulting in poor growth, yellowing, or stunting of the top portion of the plant. However, under conditions of poor soil drainage (wet soil) these fungi can girdle the crown or kill major roots resulting in plant death. These fungi are found in almost all soils. Environmental control can be achieved by improving the drainage of the soil. The fungi do not survive well in well-drained planting beds.

ROOT AND CROWN ROTS CAUSED BY OTHER FUNGI

Aside from water molds, many other fungi cause root and crown rots. The fungus *Rhizoctonia* lives in the soil and attacks a wide variety of crops. Some species of *Fusarium, Sclerotium, Cylindrocladium, Sclerotinia* and *Thielaviopsis* behave similarly. All of these fungi can persist in soil for many months through specially adapted resting structures.

Cankers and Twig Blights

Many fungi such as *Botryosphaeria, Phomopsis, Sphaeropsis* (formerly *Diplodia*), and *Fusicoccum* cause stem or twig cankers. A canker is an infection of the bark and outer vascular tissue of a tree or shrub. The infections vary in colour from the surrounding healthy bark. Infected tissue may be sunken or swollen.

When cankers girdle the twig or branch, the end of the branch dies. Small black or brown bumps, spore-forming bodies of the fungus, are often seen in the canker. These may produce an ooze of orange, white, salmon, or tan spores during rainy weather. *Phytophthora,* mentioned under root rots caused by water molds, sometimes causes cankers as well as twig and stem blight. Most notable examples include *Phomopsis* twig blight on juniper, *Dothiorella* branch blight on Shumard oak, *Sphaeropsis* tip blight on red or Austrian pine, and *Volutella* blight and canker on pachysandra. These pathogens sporulate primarily in the spring. The spores are spread by splashing water, insects, and wind.

Plant-Wilting Fungi

Verticillium is a fungus that causes wilting of a large number of woody ornamentals and some herbaceous plants such as maples, redbud, Russian olive, and chrysanthemum.

The fungus invades injured roots, grows into the stem and plugs the vascular system of the plant. Wilting, yellowing, and browning of foliage may be followed by premature defoliation.

One limb of the plant may be affected first. Gradual dieback may be apparent. Yellow-brown, brown, black, or greenish-black streaks may be found in the outer rings of wood of infected branches, or in the stems of herbaceous plants. These streaks differentiate Verticillium wilt from most other types of plant wilting and dieback. The fungus that causes Verticillium wilt has specialized "resting" structures that allow it to persist in soil for many years. *Ophiostoma ulmi* (formerly *Ceratocystis ulmi*) is the plant-wilting fungus that causes Dutch elm disease in a variety of elms. It is spread from plant to plant by root grafts or by elm bark beetles feeding on small twigs or by root grafts.

INTEGRATED MANAGEMENT OF ORNAMENTAL DISEASES

Integrated management of diseases of ornamentals should emphasize the integration of all available control methods into one program. The use of fungicides for control of several important diseases can be a major part of the overall disease management program, but the use of various cultural practices is sometimes even more important in obtaining effective disease control.

Furthermore, many important diseases cannot be directly controlled by pesticides, thus their control is almost completely dependent upon the use of cultural practices and resistant varieties. An effective disease management program should emphasize the integrated use of specific cultural practices, knowledge of the pathogen and disease biology, disease resistant plants, and timely applications of fungicides when needed. Fungicides can play an important role in this program; however, emphasis should be placed on minimizing the overall use of fungicides, while maximizing their benefits.

Identifying and Understanding the Major Diseases

It is important to be able to recognize the major diseases that occur in landscapes Proper disease identification is critical to making the correct disease management decisions. In addition to this publication, there are several other sources of useful information that can aid in developing an understanding of the major diseases of landscape plants.

The "Useful References" contains literature with colour photographs of disease symptoms on ornamentals, as well as in-depth information on pathogen biology and disease development. Many of these publications also contain excellent colour photographs and information about insect pests as well. Developing and maintaining a library of resources such as these provides a valuable tool and should be an integral portion of the disease management program.

Cultural Practices and Considerations that Aid in Disease Management

Awareness of the importance of cultural practices in disease management programs has increased greatly in recent years. Cultural practices serve as important disease control or prevention tactics in several ways. Any practice that helps create an environment within the planting that is less conducive to disease development is beneficial to plant health. Sanitation practices aimed at reducing or eliminating pathogen populations are critical for controlling infectious plant diseases. Reducing plant stress through environmental modification promotes good plant health and aids in reducing damage from some plant diseases.

It is also important to note that the effective use of fungicides within the disease management program is greatly influenced by cultural practices. Fungicides are not an alternative to the use of sound cultural practices. The following practices and considerations should be highly beneficial in preventing plant health problems in "New Landscapes."

1. Choose disease-resistant varieties with good cold hardiness and drought tolerance whenever possible. Many landscape plants have been evaluated for their resistance to specific diseases and pests in addition to their aesthetic qualities. In addition, many cultivars or varieties within a species vary greatly in their cold hardiness.
2. Select healthy nursery stock for planting. Look for any obvious signs of pests and diseases. Avoid plants with galls, torn bark, trunk wounds, and cankers. Look for a slight flare at the base of tree trunks; the trunk

flare indicates that the tree was planted and dug at the proper depth in the nursery. Avoid plants with circling roots that may eventually grow into girdling roots.

3. Choose the right plant for the location (or vice-versa!). Among the most important considerations are site exposure, soil moisture, soil drainage capability, pH, and soil texture. Knowledge of individual plant requirements are a must. Some of the very common mistakes seen in Ohio plantings include: understory plants such as the flowering dogwood in hot, dry, and sunny sites; rhododendrons and azaleas in high-pH, poorly drained clay soils; pin oak on a limestone soil; Taxus or rhododendrons on a poorly drained site; or planting anything at all under the eaves of a house where it is always very dry! While it is possible to change soil pH and improve soil texture over time, it requires a commitment to intensive management, and long-term, this will work only for smaller plants in the landscape.
4. Plant properly. Soil that is well-drained and easily worked usually may be used for planting without extensive soil preparation or amendments. However, where the investment is high, a soil test to determine the need for amendments is advised. Compacted, heavy soils will benefit from working organic material into the soil. Good drainage is essential for healthy root system development; plants struggling in poorly drained soils are targets for root-rotting fungi. If water drains from the planting hole in 24 hours, there is fair drainage. Where necessary, install drain tile or regrade the site before planting, as poor drainage is not easily rectified after planting.

   Place plants in the planting hole so the top of the original soil ball before the plant was burlapped is even with the soil grade, or at most 2 to 4 inches above Enlarging the hole in a horizontal direction outward also will aid in root establishment on compacted sites. A common practice on compacted, poorly drained sites is to place the soil ball with nearly half of the ball above the soil grade; this is not an advisable practice since the portion above the grade often heats up and dries out. A far better practice where drainage is poor is to plant landscape ornamentals in raised beds. Where this is necessary, prepare the soil in the bed, not the hole! When planting balled and burlapped plants, after placing the ball in the hole, pull back the burlap and cut it away as much as possible without disrupting the soil ball.

   Tree trunks are easily damaged at planting, and these wounds can serve as entry points for pathogenic organisms. Remove all trunk wrappiong after planting. Trunk wraps are not recommended after planting. Protect trunks by removing all strings after planting, and where staking is necessary, be certain the stake does not rub against the trunk. Lastly, remember that insufficient irrigation of new ornamental plantings is probably the most common cause of poor establishment. Maintain adequate root-ball moisture after the plants are installed, especially in artificial media. Watering during dry periods is especially important for the first 2 or 3 years after planting.

5. Mulch with caution. Keeping turf away from new plantings and applying a shallow layer (1 to 2 inches) of mulch is helpful in plant establishment. However, too much mulch can keep soils excessively wet and deprive roots of oxygen or prevent water from penetrating the soil resulting in drought stress. These conditions frequently predispose plants to root-rotting diseases. Do not mound mulch around the plant A 1-to 2-inch, flat layer of mulch around the plant is most desirable.

The following practices and considerations should be highly beneficial in preventing plant health problems in existing landscapes:

1. Practice sanitation in the landscape. Rake and remove fallen leaves from around diseased plants. The fungi that cause scab on crabapple and black spot of rose overwinter in dead leaves on the ground. Removing this source of inoculum helps greatly in disease prevention. Prune out diseased wood containing fungal or bacterial cankers. When pruning diseased material, disinfest pruning tools in 70 percent alcohol between cuts to avoid spreading pathogens. Remove and destroy badly diseased plants. Discard, burn, bury, or compost diseased plant material. Propagate from disease-free plant materials only.
2. Prevent crowding to improve air circulation in the landscape by selective removal or occasional renovation of plantings. Free water on the surface of plant tissues is required by many pathogenic fungi in order to germinate and infect the plant. Any practice that promotes faster drying time of plant tissues greatly reduces the potential for infection.
3. Where foliar diseases are potential problems, do not irrigate late in the day or at night. Water that remains on foliage overnight can promote disease development by allowing plant pathogens to become established.
4. Fertilize to promote plant vigor, but be careful not to over-fertilize. Follow normal recommendations for each plant type; too much fertilizer may be more harmful than too little. Do not fertilize stressed plants simply on the assumption that it might help; fertilize only if it is warranted.

5. Do not make flush pruning cuts or leave large stubs Make cuts so that the branch collar is not disturbed Do not paint pruning wounds. Whenever possible, remove branches when small (1-inch diameter or less). Do not "top" trees; remove large branches (greater than 4 inches in diameter) only when absolutely necessary. Pruning wounds are common entries for many plant disease-causing organisms.
6. Protect the root environment of woody ornamentals. Construction around trees and shrubs can be very harmful to plant health. Removal or addition of soil or trenching close to trees or shrubs cuts major roots resulting in extreme acute stress to plants. Avoid repeatedly digging and planting of annual flowers under the drip-lines of trees. Perennial ground covers surrounding woody ornamentals are preferred to annual plantings that require digging every season.
7. Protect the trunks of trees and other woody species. "Lawn mower" or "weed eater" blight is a very common cause of plant damage, especially on younger trees and shrubs. Damage from this type of equipment can easily "girdle" a young tree. A weed and grass-free strip around the base of a tree or shrub is very beneficial in this regard. The strip should be wide enough to allow for mowing without touching the trunk. Mulches or other ground cover materials are beneficial in maintaining a weed free strip and can be quite attractive in the landscape. In addition, bark mulches may provide some level of biological control for certain root rotting diseases.
8. Improve the root environment when necessary. Soil compaction frequently becomes a problem in established plantings. Core aeration, sometimes called vertical mulching, is used to help improve drainage and increase the oxygen levels in the soil. The goal is to improve the soil environment for root growth. Core aeration is a labour-intensive practice and is usually done to improve the health of valuable landscape specimens. The size, spacing, and depth of the holes will vary depending on the situation and the condition of the plants. As a general recommendation, auger the holes 1 to 2 inches wide, 12 to 18 inches deep. Place the holes around the drip-line of the tree, usually on 12- to 20-inch centres. Usually the holes are filled with a mixture of peat and a coarse aggregate. If a narrow diameter auger is used, holes may be left empty.

Fertilizers are sometimes placed in the holes, but since core aeration is frequently resorted to when a tree shows symptoms of decline, this should be avoided. One exception would be where iron chlorosis is a problem. Iron sulfate or iron chelate amendments may be beneficial if placed in the holes.

DISEASE-RESISTANT VARIETIES

The use of plant species or varieties that are genetically resistant to diseases is probably the most effective means of controlling plant disease. Unfortunately, many of the plants that are common in Ohio's landscapes do not have resistance to many of the more common diseases; however, when disease resistance combined with suitable horticultural characteristics is available, it should be used as an integral part of the disease management program. The species or cultivar planted often determines whether certain diseases are likely to develop or can be avoided. Resistant varieties either cannot be infected by certain pathogens or are not seriously damaged if they do become infected. Landscape plants with resistance to certain diseases. When appropriate, notes on disease resistance are provided in the section on general and specific diseases.

Check with the Ohio State University Extension office in your county or a certified nursery person for the most recent recommendations and use the information provided here to choose resistant species and varieties that are well suited to local conditions. If you select a more susceptible variety because of other preferred horticultural characteristics, be prepared to accept disease damage or devote the effort and resources required to manage it.

Crabapples and Disease Resistance

Crabapples are very popular ornamental plants for landscapes, commercial settings and parks and arboreta in Ohio. One potential problem with certain crabapples is infectious disease problems such as apple scab, fireblight, frogeye leaf spot, and occasionally cedar rusts, powdery mildew, and others. There are hundreds of crabapple species, cultivars, varieties, and selections. Following are descriptions of some of the crabapples rated highly in Ohio for both overall ornamental aesthetics and overall disease resistance in recent years*. This list is not comprehensive. (These ratings for Ohio may differ in other locations due to greater disease pressure, for example, fireblight in certain mountain states, or the presence of localized races of pathogens.)

Some Suggested Crabapples for Ohio

- *Malus baccata* (Jackii)—Maroon-red fruits. White flowers. Glossy, green foliage is exceptional ornamental feature. Handsome rounded form and structure. Grows to 30 feet tall and 25 feet wide. Excellent overall disease resistance.
- Bob White—Yellow to gold fruits. White flowers. Broadly rounded crown. Grows to 20 feet tall and 20 feet wide. Tremendous show of

delicate white flowers in spring and small golden fruits in fall and winter months. Excellent overall disease resistance.

- Donald Wyman—Bright red fruits. White flowers. Lustrous green foliage. Attractive flaking bark. Rounded tree grows to 20 feet in height and 20 feet in spread. Fruits held persistently well into the winter months. Some scab susceptibility, but not a detraction in most years.
- Mary Potter—Red fruits. White flowers. Attractive bark. Spreading-weeping habit, with ultimate height of 15 feet and width of 20 feet. Trace of scab and slight fireblight incidence, but overall good disease resistance.
- Molten Lava—Red-orange fruits. White flowers. Fruits and yellowing fall foliage provide a fiery cascading image. Weeping-spreader with multilayered overlapping branches providing fine winter effect. Height of 12 feet and width of 15-foot spread. Some scab but not a factor ornamentally in most years.
- Prairifire—Red-purple fruits. Coral-red flowers. Dark green leaves are tinged with red-purple. Flower and fruit show is outstanding. Rounded, airy growth habit with height of 20 feet and width of 20 feet. Excellent overall disease resistance.
- Red Jade—Red fruits. White flowers. Spreading, weeping habit is attractive year-round feature in the landscape. Height of 15 feet and width of 15 feet. Some scab but not ornamentally detracting in most years.
- Red Jewel—Cherry-red fruits. White flowers. Small growth habit. Fruit persists well into the winter months. Height of 15 feet and width of 12 feet. Excellent disease resistance.
- Malus sargentii—Red fruit. White flowers. Attractive low-spreading growth habit is an excellent feature, especially as plant ages. Very effective in group plantings. Height of 8 feet and width of 16 feet. Excellent disease resistance.
- Sentinel—Red fruits. White with pink-tinged flowers. Pink buds opening to white flowers are sensational. Vase-shaped upright habit. Height of 15 feet and width of 10 feet. Trace of scab but not detracting ornamentally.
- Strawberry Parfait—Fruits mostly red with yellow base. Pink flowers with red margins. Unusual erratic upright-spreading growth habit is one of its best features. Height of 20 feet and width of 25 feet. Fruits remain firm into the winter months. Foliage is striking as it develops along the upright branches in the spring. Excellent disease resistance.

- Sugar Tyme—Brilliant red fruits. White flowers. Broadly rounded growth habit with a height of 18 feet and width of 15 feet. Prolific, showy fruits persist long into the winter months. Trace of scab but not detracting ornamentally in most years.

Using Fungicides

Fungicides can be an important component of the disease management program. However, it is important to remember that their use should be integrated with the use of sound cultural practices, a knowledge of pathogen and disease biology, and disease-resistant or less susceptible plants whenever possible. Fungicides are only effective when infectious plant diseases that are caused by fungi are truly the cause of the problem. In some cases, pests and diseases follow other environmental imbalances and may not be the major problem. In cases such as these, the fungicide will help but is often not the total answer. Also, it is important to remember that fungicides are only effective if several rules are followed. First, the correct material must be selected. This depends on correct diagnosis and identification of the pathogen. Second, the chemical must be applied at the right time of year and frequently enough to protect plant material adequately. Third, fungicides must be applied properly over plant surfaces. These three rules depend on your making correct decisions based on correct knowledge. Too many people simply "spray and pray," and are often disappointed with the results.

CONSIDERATIONS FOR ROTATION TO BE AN EFFECTIVE DISEASE MANAGEMENT TOOL

For rotation to be an effective strategy for disease management producers need to keep in mind that certain requirements or conditions must be met for this practice to be effective. Firstly, the disease source must be from the field itself or where there is limited potential for movement of plant pathogens from adjacent fields. Consequently, crop rotation will tend to be most effective for those diseases that are soil-or residue-borne. Crop rotation will not be effective for diseases where seed-borne pathogens are an important source of disease, such as the cereal smuts. Furthermore, rotation will not be effective where pathogens are mobile and readily move from one field to another or over long distances.

Secondly, the host range of the pathogen needs to be narrow. Crop rotation may not be an effective strategy for a disease like sclerotinia stem rot of canola because the pathogen that causes this disease has an extensive host range including most common broad-leafed weeds and crops, and thus would have ample opportunity to persist over growing seasons via other potential host plants. The final condition for rotation to be effective would be that pathogens are unable to survive for long periods in the absence of a suitable host. Thus, there should be a rapid reduction in pathogen viability and its capacity to produce infective structures, i.e. spores, within a relatively short period of time. Rotation will not be as effective for those pathogens that have the capability of producing long-lived resting structures or where they can survive in infested plant tissues that resist decay. *Sclerotinia sclerotiorum* produces resting structures known as sclerotia that are fairly long-lived. The blackleg fungus can survive in old canola root pieces for extended periods of time since these plant tissues are fairly resistant to decay and can persist for periods of three or more years.

Let us look at some examples where rotation can have a beneficial influence on disease management, where rotation may need to be combined with other strategies, and a bit of a spin on the traditional concept of crop rotation.

THE ROTATIONAL BENEFIT OF PULSE CROPS

One feature of many ancient crop rotations was the inclusion of legumes including pea, bean, vetch, and lupins. The benefit of including legumes in crop rotations was recognized in ancient times and is also recognized today. Crop rotations that include legumes have the ability to provide succeeding crops with nitrogen, reduce disease incidence and improve weed control strategies. Inclusion of legumes in rotation will cause changes in soil fertility, soil microorganisms, soil organic matter, soil water and crop responses (Biederbeck et al. 1995). The focus of this section will be the rotational benefit of pulses with regard to root disease management in subsequent cereal crops.

Research conducted in Australia over the past twenty years has illustrated the benefit of pulse crops for subsequent cereal crops. Reeves et al. (1984) studied the effect of lupin on subsequent wheat crops from 1974 to 1979. Inclusion of lupin either in the preceding year or even two years previous, significantly reduced the level of take-all, caused by the fungus*Gaeumannomyces graminis* var. *tritici,* and increased yield compared with a continuous wheat rotation. Doyle et al. (1988) found that the incidence and severity of common root rot caused by the fungus *Cochliobolus sativus* was reduced in wheat following lupin compared with wheat following wheat. Yields were also significantly higher in wheat following lupin and this increase was attributed to suppression of disease and better nitrogen fertility.

Rotational benefits with regard to disease management have not been confined to lupin. Cotterill & Sivasithamparam (1988) also found a reduction in take-all after rotation to field pea, lupin or oats. Recently, Felton et al. (1998) demonstrated that crown root in wheat, caused by the fungus *Fusarium graminearum* Schwabe, was reduced and wheat yields increased in a chickpea-wheat versus a wheat-wheat rotation. Overall, mean incidence of crown rot was 12% for the chickpea-wheat rotation compared with 30% with the wheat-wheat rotation. Felton et al. (1998) attributed the increase in yield with wheat preceded by chickpea to enhanced nitrogen fertility and a reduction in disease. In western Canada, Stevenson & van Kessel (1996a) found a significant reduction in common root rot severity of wheat with a pea-wheat versus a wheat-wheat rotation at one of two sites in Saskatchewan. In addition, Stevenson & van Kessel (1996b) in a landscape-scale study found a slight reduction in common root rot severity with the pea-wheat rotation, but this reduction was not significant. Bailey et al. (1992), also found reduced levels of common root rot when spring wheat was preceded by field pea or fallow compared with spring or winter wheat.

Research conducted in the United Kingdom also illustrated the disease management benefit of legumes on subsequent barley crops. Dyke & Slope (1978) found that over a range of nitrogen rates (0 to 113 kg/ha) take-all was severe in barley after barley, but was a minor problem in barley after oats, beans or red clover. Yields were reduced in barley after barley compared with barley after oats, beans or red clover at all nitrogen rates.

At Fort Vermilion, Alberta we have been involved with a crop management field trial evaluating conventional versus zero tillage systems and legume-based crop rotations. The trial was originally established in 1992. Precrop treatments for this study included red clover as a green manure, field pea as a seed crop, summerfallow and continuous wheat. The crop rotations included: 1) field pea-wheat 1-canola-wheat 2 (FP/W1/C/W2), 2) red clover-wheat 1-canola-wheat 2 (RC/W1/C/W2), 3) fallow-wheat 1-canola-wheat 2 (F/W1/C/W2), and 4) wheat-wheat 1-wheat-wheat 2 (W/W1/W/W2). Common root rot (CRR) severity in the wheat was assessed in early August of 1994, 1995 and 1996. Overall, CRR severity was reduced under zero versus conventional tillage. Root rot severity was also reduced in rotations with field pea, red clover or canola as precrop treatments compared with continuous wheat. In addition, CRR severity was reduced in legume-based rotations compared with non-legume rotations.

CROP ROTATION AND CEREAL LEAF DISEASES

In a study of commercial barley fields in Alberta from 1995 to 1997, there was a tendency for lower levels of scald and net blotch to occur when barley was planted in fields previously cropped to a non-host. For scald, a rotation of 1-2 years away from a susceptible host maybe sufficient to allow for the destruction of most infected plant tissue, since the scald pathogen primarily infects leaves and leaf sheaths, which decompose fairly rapidly. However, a longer rotation will be needed for net blotch, since the fungal pathogen that causes this disease also infects barley stems, which are more resistant to decay and will persist, especially the node tissue, for a longer period of time. Results from the Alberta barley leaf disease study also indicated that extended rotations might not always reduce a producer's risk of scald and net blotch. Substantial levels of both scald and net blotch developed in fields not planted to barley or forage grasses for the previous four years. Average percent leaf area diseased (PLAD) of the flag leaf-1 in each year ranged from 0.9 to 3.2% and 3.5 to 9.9% for scald and net blotch, respectively. However, in some individual fields scald and net blotch levels on the flag leaf-1 exceeded 30 and 50% PLAD, respectively. Most of the high levels of disease, especially for net blotch occurred in fields of Harrington. One potential source of disease in these fields may have been seed-borne inoculum, especially for the causal agent of scald, which is a rain-splash dispersed pathogen with limited potential for field-to-field movement of disease. The causal agent of net blotch is a fungus that produces wind-borne spores and thus, will have more potential for field-to-field movement.

In Alberta, Skoropad (1959), Lee et al. (1999), and Xi and Burnett (1997) found that *Rhynchosporium secalis*, the causal agent of scald, infected barley awns, lemmas, paleas, and pericarps. In the 1950's, Skoropad (1959) found that seed samples collected randomly from Alberta barley crops had from 1 to 10% scald infected seed with an average of 2.0%. More recently, during the summer of 1996, seed from a regional variety trial located at Westlock, Alberta was found to have scald infection levels that ranged from 0 to 10% infected seed.

The movement of scald into scald-free areas via infected seed has been suggested by a number of researchers. Research in the 1950's and 1970's demonstrated transmission of *R. secalis* from infected seed to seedlings mainly under greenhouse conditions. Reed (1957) also observed that infected seed that remained on the soil surface after seeding could act as a source of inoculum. In some countries, including India, New Zealand and England seed-borne inoculum of net blotch is considered to be an important source of disease. At Lacombe Alberta, Piening (1968) studied seed to seedling transmission using seed that was covered with cultured net blotch mycelium or ground net blotch infected leaves. In the field, seedling infection levels were 0.5 to 1.5% and 0 to 0.15% for infested and control seed, respectively.

At Indian Head, Saskatchewan, Bailey et al. (1992) found that leaf spot severity was higher in wheat following spring or winter wheat compared with field peas or summerfallow. Pedersen & Hughes (1992) found in the Parkland area of Saskatchewan, that when environmental conditions were favourable for disease development Stagonospora/Septoria leaf spot (*Stagonospora nodorum*) severity was reduced in wheat by a rotation with two years between wheat crops, but not for rotations with only one year between wheat crops. However, when environmental conditions were less conducive to disease, a rotation with one year out of wheat reduced Stagonospora/Septoria leaf spot severity to the same extent as a rotation with two years between wheat crops. In southwestern Saskatchewan, Fernandez et al. (1998) found in a research trial and in commercial wheat fields that leaf spot severity was not always lower in wheat planted after a non-host compared with continuous wheat. They concluded that at least two years out of wheat were required for decomposition of infested residues to reduce wheat leaf spot levels. In 1996 and 1997, a wheat leaf disease survey was conducted in commercial fields in Alberta. In both years, similar leaf spot levels (septoria complex and tan spot) were found in wheat grown on wheat and wheat grown on a non-host. Furthermore, leaf spot levels in wheat fields cropped to non-hosts for the previous two to four years tended to be similar to wheat planted on wheat residue.

Research from the United States has indicated that seed infection in wheat may be a factor that needs to be considered when using rotation as a means of cereal leaf spot management. In Florida, Luke et al. (1983) found decreased development of Stagonospora/Septoria blotch, caused by *S. nodorum*, with either a 1 or 2 year rotation using uninfected seed, but not when *S. nodorum* infected seed was used. Similar results were also found by Milus & Chalkley (1997) in Arkansas where low (7%) and high (34-40%) levels of seed infection with *S. nodorum* led to significant leaf infection in an experiment planted in an area not cropped to wheat for at least the previous four years. They concluded that crop rotations should be combined with appropriate seed treatments to help prevent outbreaks of Stagonospora/Septoria blotch. In New York State, Shah et al. (1995) found significant relationships between *S. nodorum* seed infection levels of up to 40% and subsequent development of Stagonospora/Septoria blotch in winter wheat. Seed to seedling transmission of *Pyrenophora tritici-repentis*, the causal agent of tan spot of wheat, has been demonstrated under greenhouse conditions, and under laboratory conditions and in artificial potting media under outside conditions.

At present little evidence appears to exist, especially in wheat, concerning the potential role of seed-borne pathogens as a source of inoculum for cereal leaf spot epidemics in western Canada. Recent seed surveys have shown that over a three year period from 1995 to 1997 between 71 and 97% of barley grain samples in Alberta had detectable levels of infection with *Pyrenophora teres*, the causal agent of net blotch. Samples from 1995, 1996, and 1997 had overall average seed infection levels of 7.1, 12.1, and 22.6%, respectively. The maximum observed seed infection levels were 82, 81, and 89%, for 1995, 1996, and 1997, respectively.

The frequency and level of infection tended to be higher for 2-row barley samples compared with 6-row samples. Based on results from Edney and Tipples (1997) the majority of 2-row samples would likely be the variety Harrington. The same survey showed that over a three year period from 1995 to 1997 between 39 and 75%, and 92 and 100% of wheat grain samples in Alberta had detectable levels of infection with *P. tritici-repentis*, and *S. nodorum*, respectively.

Samples from 1995, 1996, and 1997 had overall average seed infection levels with *P. tritici-repentis* of 0.6, 2.0, and 2.1%, respectively, while yearly average infection levels with *S. nodorum* were 22.3, 16.9, and 16.7, respectively. The maximum observed seed infection levels with *P. tritici-repentis* were 3, 9, and 12%, for 1995, 1996, and 1997, respectively, while maximum levels of seed infection with *S. nodorum* were 43, 54, and 54%, respectively.

Barley Variety Rotation, A Spin on the Traditional Idea of Crop Rotation

Currently, a significant proportion of the diet utilized for beef production in Alberta is barley silage and feed grain. Alberta accounts for close to 50% of the total barley acreage in western Canada. Barley is an attractive feedstock as a result of its high yield, protein and digestibility and because farmers are familiar with its production, have ready access to seed and have a choice of seed or silage production. An estimated 48% of the barley grown in Alberta is for feed purposes and in some regions this can be as high as 60-70%. To ensure a constant supply of feed material many farmers will often grow barley continuously for several years.

For example, in a survey conducted by McLelland (1992) an average of 36% of the producers grew barley on barley with the proportion increasing to 75% in the western parts of central and southern Alberta. However, continuous barley production can result in significant disease problems, especially where the same variety is grown year after year. Where a continuous supply of feed grain is needed and potential alternatives are lacking, rotation to non-hosts as a management tool for leaf diseases of barley is not an option.

At Lacombe, we are looking at the potential of rotating barley varieties with different genes for resistance to diseases like scald as a means of reducing the impact of leaf spot diseases. Rotating barley varieties with different genes for resistance would be analogous to what producers are currently doing to prevent the build up of herbicide resistant weeds by rotating herbicides with different modes of action.

Different barley varieties may have different sources of disease resistance, especially if they are from different breeding programs. Growing varieties with different sources of resistance and perhaps combined with rotation to suitable non-hosts may help to prevent the selection and build-up of a specific pathogen race, and thus prolong the usefulness of a particular source of resistance.

In 1998, a 3-year experiment was set up at Agriculture and Agri-Food Canada, Lacombe to investigate the potential of barley variety rotation as a means of reducing the impact of leaf diseases and maintaining crop productivity. Preliminary results will be presented from the trial at Lacombe.

3

Plant Diseases Management

The interactions between plants and disease organisms are complex, and commercial growers and home gardeners alike may have difficulty understanding plant diseases. Confusion can be reduced by learning a few basic concepts and principles of how diseases develop and how they are managed.

This publication presents these concepts and is dedicated to commercial growers, commercial applicators, home gardeners and others who want more information on how plant diseases develop and are managed.

To serve these diverse groups, examples of common North Dakota diseases were selected from both the commercial farm and from the home garden. Various circulars available at offices of the NDSU Extension Service provide detailed information on many of these diseases.

HOW DISEASE DEVELOPS

What is a Disease?

A disease is any abnormal condition that damages a plant and reduces its productivity or usefulness to man. Under this definition, air pollution can cause disease, as can many fungi and other infectious living organisms. This illustrates the first important concept: there are two basic types of diseases, non-infectious (abiotic) and infectious (biotic).

TYPES OF DISEASES

Non-infectious (Abiotic)

Non-infectious diseases are caused by some environmental factor that produces an abnormal plant; that is, one that has an abnormal appearance. Non-infectious diseases are *not* caused by a living, parasitic organism (an organism that gets its food by attacking other organisms), but are abiotic in nature.

Nutrition

Nutrition is a frequent cause of non-infectious disease. Either too much (excess) or too little (deficiency) can cause problems. For example, plants that are deficient in nitrogen develop a general yellowing, beginning with the lower leaves and progressing upward. Trace element deficiencies such as iron chlorosis, caused by iron deficiency, are common. Iron chlorosis occurs in many North Dakota trees and shrubs, especially silver maple, oak, and spirea. Iron chlorosis is recognized by progressively smaller leaves on the new growth; these leaves are yellow with green veins. When iron chlorosis is severe, leaves may turn brown and become brittle as well. Lime-induced chlorosis is common in our alkaline soils because the iron in the soil is not readily available to plants. Iron chlorosis also is common on certain soybean varieties. Zinc deficiency is common on dry beans and fairly common on flax, causing yellow leaves and stunted growth. Excess trace elements may also cause growth problems, but these are rare in North Dakota.

Moisture

Deficient or excessive moisture (water) can cause disease. Moisture deficiency produces stunted, stressed or wilted plants. In addition, this stress may predispose (weaken) plants to infection by infectious organisms or increase the effects of infectious disease. For example, some tree canker organisms commonly infect trees stressed by drought or extreme cold. The effects of stem rust and root rot on small grains are greater when plants are moisture stressed (deficient in water). Excess moisture also has adverse effects, such as suffocation of roots due to lack of oxygen or predisposing plants to water mold infections.

Temperature

Frost is a common problem in spring and fall, affecting tender farm crops and garden vegetables. Extremely high temperatures in summer can also cause problems. For example, heat sterility in small grains is common in North Dakota. In oats, this is referred to as "blast."

Other Meteorological Conditions

High soil temperatures early in the season may injure or kill plant tissues at the soil surface, resulting in a constricted stem; this is called heat canker. Bright sun, high temperatures, and strong dry winds may suddenly desiccate (dry) leaves of crops and garden plants, resulting in sunscald. When lightning strikes the ground it may kill plants in somewhat circular patches up to 50 feet in diameter.

Toxic Chemicals

Toxic chemicals injure plants. Salt may damage or kill farm crops growing in saline seeps; road salt may severely damage boulevard trees and other vegetation. Air pollution also damages vegetation. Bronzing of beans caused by ozone is common in the state. The source of the ozone is not known.

INFECTIOUS (BIOTIC)

Infectious diseases are caused by organisms that attack plants and get their nutrition from them. The plant attacked is called the host plant. The organism causing the disease is called a pathogen. The pathogen can spread from a diseased plant to a healthy plant. There are five common groups of pathogens A few other kinds of microorganisms may cause plant disease but are not common in North Dakota.

Fungi

Fungi are the most common pathogens in North Dakota. They produce tiny thread-like filaments called hyphae. Most pathogenic fungi produce spores which serve to reproduce and disseminate them. Spores function similarly to the seeds of higher plants. Some spores are formed in masses, like the orange pustules of rust fungi. Other spores develop in specialized fruiting structures. These structures are called signs of the pathogen and are useful in field identification of disease. Symptoms are also useful in identification of a disease. Symptoms are visible abnormalities such as wilts, rots, and other types of tissue death, stunting, excessive growth, or abnormal colour.

Examples of common pathogenic fungal diseases in North Dakota include: rusts of small grains, sunflower and dry beans; cereal smuts and head scab of small grains; Cercospora leafspot of sugarbeet; white mold or Sclerotinia of dry beans, sunflowers, and canola; early blight and late blight of potatoes; root rots of small grains, sugarbeets and dry beans; apple scab; anthracnose of muskmelon; tree cankers; Septoria leafspot of tomato; peony blight; powdery mildews of ornamentals, and plum pockets.

Most fungi that cause plant diseases are parasites, organisms that get their food from other living organisms. However, not all fungi are parasites. Many live on dead or decaying organic matter and are called saprophytes. Mushrooms that spring up in lawns are among the most spectacular saprophytic fungi. There are also many inconspicuous ones that rot organic matter. The sooty molds seen on wheat heads at harvest also are saprophytic, living on the already ripe or senescing glumes and awns.

Bacteria

Bacteria are tiny one-celled organisms that multiply by cell division. They can be seen only with a microscope. Most are saprophytes, but there are a few common and serious bacterial pathogens that attack North Dakota plants.

Examples of common bacterial diseases include bacterial blights of dry beans, bacterial blights and black chaff of wheat and barley, ring rot and blackleg of potato, fireblight of apples and related plants, bacterial wilt of cucumber and muskmelon, angular leafspot of cucumber, and bacterial speck and spot of tomato.

Viruses

Viruses are 1,000 times smaller than the tiniest living cell. Most viruses have a core of nucleic acid, the basic unit of heredity, and have a protein coat covering the core. Viruses are usually in the form of rods or spheres and alter the activities of the host to manufacture more virus.

Some viruses are transmitted mechanically (by contact with another plant, or contaminated workers' hands or tools); others are transmitted (carried) by insects and by eriophyid mites. Examples of virus diseases that can cause serious losses are wheat streak mosaic, barley yellow dwarf, bean common mosaic, potato virus diseases, tobacco mosaic, cucumber mosaic, and squash mosaic.

Phytoplasmas

Phytoplasmas lack a rigid cell wall, have no defined shape and can only be seen with an electron microscope. They are usually systemic in the host (distributed internally throughout the host) and are transmitted by leafhoppers.

Phytoplasmas cause growth abnormalities such as witches' brooms (a broom-like mass of plant branches) or excessive tillering (stooling of small grains).

Our most common phytoplasma is aster yellows. Aster yellows produces witches' brooms and greenish flowers on marigold; carrots develop yellow tops and hairy roots. The disease produces greenish flowers on flax, "purple top" on potato and tomato, and bladder-like pods on canola. Purple top of potato is often accompanied by the formation of small aerial tubers in the leaf axils (the point where the leaf joins the stem).

Nematodes

Nematodes are tiny roundworms. Most can be seen only with a microscope, but a few can be seen with the naked eye. Reproduction is by formation of eggs. Some parasitic forms attack plant roots and can cause severe damage. Nematode problems are common in warm climates but are rare in North Dakota. Two potentially serious nematode diseases occur in neighboring states but are not yet confirmed (2000) in North Dakota. They are the soybean cyst nematode and the pine wilt nematode. The pine wilt nematode is unusual in that it invades the vascular (water-conducting) tissues of pine trees instead of attacking the roots.

DISEASE DEVELOPMENT

This section discusses how plant pathogens produce disease: how they infect the host, how they reproduce, how they are disseminated (spread), and how they survive between crops.

What Does it Take to Produce Disease?

Three factors interact to produce disease; the host, the pathogen, and the environment If any one of these three factors is unfavourable or missing, disease will not develop. For example, the flax rust pathogn attacks only flax as a host, and different races of flax rust attack different varieties of flax. Flax rust develops only when a suitable combination of rust race (pathogen) and variety (host) interact under environmental conditions favouring this disease. The environment often limits disease in North Dakota and is the reason that some diseases rarely occur and others occur sporadically. Many diseases are favoured by humid or rainy weather and may be more common and severe in years favouring good crop production.

FUNGICIDES

Protectant fungicides act on the plant surface to protect against infection, and systemic fungicides are taken up by the plant tissues and then function to prevent infections. Some new fungicides have limited therapeutic (curative) properties.

Protectant Fungicides

Protectant fungicides work on the plant surface to prevent spore germination or kill developing fungus hyphae before the host plant is penetrated and infection becomes established.

Protectant fungicides commonly used on North Dakota farms include maneb (Maneb 80, Maneb 75 DF), mancozeb (Dithane DF, Manzate 200 DF, Penncozeb DF), triphenyl tin hydroxide (Super Tin, Agri Tin), chlorothalonil, and copper fungicides. Protectant fungicides commonly used on North Dakota home gardens or ornamentals include captan (Orthocide), chlorothalonil, maneb, mancozeb, thiram, and copper fungicides. The dicarboximide fungicides iprodione and vinclozolin also are protectant fungicides that provide good control of Sclerotinia (white mold) on certain field crops.

Most protectant fungicides cannot stop development of a pathogen once infection occurs. Leafspots become visible a number of days after infection: four to five days for tan spot of wheat, potato late blight and sugarbeet Cercospora leafspot, and seven to 14 days for many other pathogens. If a protectant fungicide is applied after infection has occurred but before leafspots show, the disease is not cured but continues developing and symptoms appear. Then frustrated growers and gardeners may claim that the fungicide "didn't work." Protectant fungicides do not cure infections and will not work unless the application is timely (early enough). Waiting too long usually results in failure. A few examples will illustrate this.

Gardeners sometimes begin spraying their tomato plants with a protectant fungicide such as Ortho Multi-Purpose Fungicide when all the lower leaves on their tomato plants are dead and the middle leaves already are heavily spotted with *Septoria* leaf spot. The disease continues to develop on the middle leaves resulting in defoliation (loss of leaves). Since some upper leaves were already infected before spraying started, symptoms may develop on these leaves, too. Only two or three tip leaves were uninfected at the first spraying. These leaves were all that was protected against infection and are all that will be left after disease has defoliated the rest of the plant. If the gardener had sprayed when the first few spots showed on the lower leaves, some lower leaves might have been lost to *Septoria,* but the rest of the plant would have remained healthy.

Timely application of protectant fungicides also is critical for crop diseases, such as rust of dry beans or leaf rust of winter wheat. Rust diseases have a tremendous capacity to reproduce in a short period of time. Rapid development of these diseases is favoured by very susceptible varieties and rainy, humid weather.

If rust is detected on these susceptible crops in a given area, if environmental conditions are favourable, and if those crops have good yield potential, growers should apply protectant fungicides to the crop prior to rust formation on the healthy leaves. Applying too late will mean that rust infections will already have occurred and new spore formation will not be stopped. Protectant fungicides are often combined with a spreader-sticker which reduces wash-off of the fungicide during rainfall. These compounds help to wet the leaf and to bind the fungicide to the leaf. Protectant fungicides must be applied so that thorough coverage of the foliage is obtained. Protectant fungicides are redistributed on the plant by dew and rain, but uniform coverage is still essential.

A fungicide should be applied before a rainy period to provide protection during the rainy period. The application must be made early enough so that the spray droplets have dried before the rain begins. If necessary, another application can be made after the rains are over.

Sulfur is a protectant fungicide that effectively reduces powdery mildew on a number of crops, including sugarbeets, garden and dry peas and ornamentals. Sulfur also has some eradicant action against established powdery mildew infections. This therapeutic effect of sulfur is an exception to the general rule that protectant fungicides do not cure infections.

Systemic Fungicides

Systemic fungicides are taken up (absorbed) by the plant. They include some seed treatment fungicides such as carboxin, the benzimidazole fungicides, the sterol inhibitors, the phenylamides and the strobilurins. Benzimidazoles are commonly used as foliar fungicides and include benomyl, thiabendazole, and thiophanate methyl (Topsin M). These fungicides move upward in the plant but cannot move down the leaf or stem. They function as locally systemic fungicides and usually do not move into new foliage. Most sterol inhibitors and strobilurins are locally systemic.

Locally systemic fungicides have several advantages: they protect both sides of the leaf even if only one side was sprayed, they are not washed off by rain or decomposed by sunlight, the interval between sprays may be longer than that with the protectant fungicides, and some have a therapeutic effect. The therapeutic effect ranges from 24-36 hours for the benzimidazoles and up to four days for some sterol inhibitors.

Benzimidazole Fungicides

Benomyl and thiophanate methyl are currently registered for control of white mold *(Sclerotinia)* on dry beans and give good control if properly applied at the right time. The white mold fungus infects dry beans by means of airborne spores which must begin growth on dead plant tissue, and then growth spreads to green tissue. The site of this dead tissue is the dead blossoms or dead lower leaves. Since these fungicides cannot move down the plant, they are effective only if there is complete coverage of the entire plant. Canopy penetration is essential to good control.

Phenylamide (Acylalanine) Fungicides

Metalaxyl (Allegiance) and mefenoxam are systemic fungicides with activity against water molds and related fungi such as *Pythium, Phytophthora* (potato late blight pathogen) and the downy mildew organisms. Oxadixyl (Anchor) is a systemic seed treatment fungicide with similar mode of action to metalaxyl and mefenoxam.

Sterol Inhibitor (Ergosterol Biosynthesis Inhibitor) Fungicides

The sterol inhibitor fungicides inhibit the formation of ergosterol in the higher fungi, but not the water molds. This prevents cell wall formation and stops growth of the fungus. Sterol-inhibiting fungicides currently available in North Dakota (2000) include imazalil (several brand names), triforine (Funginex, for home garden use), difenoconazole (Dividend seed treatment), triadimefon (Bayleton) triadimenol (Baytan), propiconazole (Tilt), tebuconazole (registered Raxil seed treatments, registration pending for Folicur foliar treatments), and tetraconazole (registration pending for Eminent). They have therapeutic activity, many for up to four days after infection has initiated, may suppress spore formation in established infections, most are taken up by the plant (some are only locally systemic), and they are used at very low dosages.

Strobilurin Fungicides

The strobilurin fungicides were developed recently. This is a methoxyacrylate class of chemistry which is related to naturally occurring products found in a group of forest mushrooms called "pine-cone mushrooms." Some are locally systemic and many have an extremely broad spectrum of activity against all four classes of fungi, something that is extremely rare with most classes of fungicides.

These fungicides are environmentally friendly. Azoxystrobin (Quadris) and kresoxim-methyl (Sovran, Cygnus) were the first of this class. Other strobilurins include trifloxystrobin (Flint), BAS 500 (Headline, expected to be registered for the 2002 season) and a product with a related mode of action called famoxadone (Famoxate, registration expected in several years). Some strobilurins such as azoxystrobin are locally systemic, moving into and up the leaf toward the tip. Other strobilurins such as kresoxim-methyl and trifloxystrobin have only slight uptake by the leaf and are redistributed on the plant through vapor action. Most fungicides in this class are strongly bound to the cuticle (leaf surface) and so are not easily washed off the leaf. They inhibit fungal respiration, which is a different mode of action from that of other fungicides.

New Classes of Fungicides

Several other fungicides have been developed recently, each of which is in a unique class of chemistry. Hymexazol (Tachigaren) is an isoxazole class of chemistry that is environmentally friendly and has excellent activity against *Aphanomyces* and also *Pythium*. It is registered for use as a seed pelleting for sugarbeets for early season control of Aphanomyces seedling disease.

Another new chemistry is fludioxanil (Maxim), a phenylpyrrole class of chemistry that is an environmentally friendly product registered for seed treatment on many crops and also as a potato seed piece treatment. It has excellent activity against silver scurf of potato, as well as Rhizoctonia black scurf of potato and Fusarium dry rot. Fluazinam (Omega) is a pyridinamide class of chemistry that is a contact fungicide with a broad spectrum of activity, including activity against Sclerotinia (white mold). Fenhexamid is a hydroxyanilide class of chemistry that is used as a protectant fungicide.

Some fungicides that are in varying classes not listed above include the registered fungicides cymoxanil, dimethomorph and propamocarb and the registration pending product zoxamide All four have good activity against the potato late blight pathogen and the downy mildews. Cymoxanil is an acetimide fungicide that is locally systemic, has a short residual and has both curative and protectant properties.

Dimethomorph is a cinnamic acid derivative that is locally systemic and has curative and preventive activity. Propamocarb is a carbamate fungicide that is systemic and is translocated into new foliage. Zoxamide is an amide fungicide with a different mode of action from other potato late blight fungicides.

Some products such as acibenzolar (Actigard) and several others act to enhance the natural defenses of the plant. Currently there are no uses registered for North Dakota crops.

Resistance to Fungicides

Fungi may develop resistance to fungicides. Repeated use of the same fungicide or fungicides with the same mode of action can lead to development of resistance. Fungi are less likely to develop resistance to the older protectant fungicides that have multiple sites of action against fungi. Nevertheless, the

Cercospora leafspot pathogen of sugarbeet has developed tolerance to triphenyltin hydroxide. This tolerance is expressed as a reduced sensitivity to the fungicide, requiring higher application rates to control Cercospora leafspot.

Fungi are more likely to develop resistance to systemic fungicides with a single specific mode of action. Resistance to the benzimidazoles and the phenylamides can occur rapidly, with resistant strains of fungi showing insensitivity to high rates of the fungicide. Resistance also can develop rapidly to the dicarboximides, the sterol inhibitors, and the strobilurins.

The Cercospora leafspot pathogen of sugarbeet is resistant to benzimidazole fungicides in many areas of the world, including many fields in Michigan, Minnesota and North Dakota. The Fusarium dry rot pathogen and the silver scurf pathogen of potato have developed resistance to this class of fungicide across much of the United States and Canada. These three fungicides are closely related, and a pathogen that develops resistance to one of the benzimidazole fungicides will be resistant to all three. This is called cross resistance.

The potato late blight fungus can develop resistance rapidly to the phenylamides metalaxyl and mefenoxam. Certain genotypes of the fungus have resistance, including the US8 genotype now present in North Dakota and many other parts of the United States and Canada. The presence of this new genotype has seriously limited the usefulness of this class of fungicide for potato late blight management. The gray mold fungus has rapidly developed resistance to the dicarboximide fungicides iprodione and vinclozolin (Ronilan) under greenhouse conditions. Certain grape pathogens have rapidly developed resistance to this class of fungicide in California.

Cases of resistance to sterol inhibitors are fairly common, and resistance management practices are recommended for this class of fungicide. Although the strobilurin fungicides have been used but a few years, it is recognized that resistance can occur to this class of fungicide, and cases of resistance are already being reported worldwide.

Resistance Management

There are several ways to retard the development of resistance. These include:

- *Tank mix with a fungicide with a different mode of action.* Mancozeb or chlorothalonil can be tank mixed effectively with benzimidazole or phenylamide fungicides.

- *Alternate applications between or among two or more classes of fungicides with different modes of action.* This is a good strategy for resistance management of triphenyltin hydroxide, the sterol inhibitors, the dicarboximides and the strobilurins. Although the sterol inhibitors and the phenylamides have some post-infection activity, they are best used in a preventive manner, which reduces the likelihood that resistance will develop.
- *Apply a limited number of applications in a block at a critical period in the pathogen disease cycle.* A different mode of action should be used at other less critical times in the disease cycle, so as to minimize the exposure of the "at risk" fungicide. This has been recommended with some of the strobilurins.
- *Limit the number of applications of an "at risk" fungicide per year.* This has been done with the phenylamide fungicides, the sterol inhibitors, and the strobilurins. Use of these fungicides may be restricted to the most critical parts of the season.
- *Avoid reduced rates of fungicides.* These reduced rates may facilitate the development of resistance in fungi.
- *Do not use phenylamides as soil treatments against airborne pathogens.*

Use of Fungicides

All pesticides are poisons, and they should be used with care. Most commonly used fungicides have a relatively low level of toxicity, but triphenyltin hydroxide (Super Tin, Agri Tin) is an exception in that it is quite toxic. Many fungicides have information on the label restricting the number of days before harvest that the last application can be made to food crops. Most have a 24-hour period during which re-entry in a field should be done only if wearing protective clothing (long pants, long-sleeved shirt, shoes, chemical-resistant gloves).

Most fungicides can be stored successfully from year to year, but a dry storage area is essential, especially for the maneb and mancozeb fungicides. Care should be taken in storing liquid or flowable fungicides as some will not stand freezing.

NONCHEMICAL DISEASE CONTROL

With the increasing concern about use and misuse of pesticides in commercial agriculture and home gardens, there are more and more inquiries for organically grown commodities every year. Nonchemical control practices for plant diseases have been known and recommended for years. The backbone of any integrated pest control program must always include cultural and sanitation practices, two important components of nonchemical disease control.

Unfortunately, disease problems may begin as soon as seeds are planted and can continue into harvest and storage. Plant diseases may be caused by several pathogenic organisms, such as fungi, bacteria, viruses, mycoplasmas and nematodes. In addition, nonliving factors, such as deficiencies or excesses of water, light, temperature, air pollution, pesticides and nutrients, can either predispose a plant to disease or directly cause plant injury.

Fortunately, many disease problems can be prevented or controlled without pesticides. Effective plant disease control begins at the onset of disease or even before symptoms appear.

Resistance

Effective disease control through resistance (a plant's tolerance or immunity to a disease) is based on knowledge of the diseases that occur in an area. Always choose varieties of plants that are adapted to Colorado growing conditions. Many vegetable, fruit and ornamental plant varieties are available with resistance to one or more diseases.

For example, when purchasing tomato varieties, always select plants labelled "VFN," "VFNA," "VFNT," etc. This indicates that the plants are resistant to Verticillium wilt (V), Fusarium wilt (F), southern root-knot nematode (N), early blight (A), or tobacco (tomato) mosaic virus (T). Selecting resistant plants may eliminate many disease problems. Contact your Colorado State University Extension county office for lists of plant varieties that are successfully grown in this area.

Exclusion

Exclusion is preventing the entrance and establishment of disease-causing organisms (pathogens) into areas where plants are grown. This means avoid bringing diseases into the garden or moving them around within the garden.

Use certified, disease-free seed or transplants. Examine the leaves and root systems of transplants and eliminate or destroy diseased plants. Either raise your own transplants in sterilized beds or buy them from a reputable dealer. Do not purchase transplants with galls or swellings on their roots or plants that have a brown discoloration on the stem at the ground line. Galls or swellings may indicate root-knot nematode infection. A brown stem discoloration may mean the presence of damping-off organisms. Also, avoid transporting soil or tools from known disease areas to disease-free areas.

Eradication

Eradication is the elimination of the disease-causing organism after it has become established on a plant. Eradication can be accomplished by several methods.

Sanitation

Plant pathogens are less likely to survive if organic matter is quickly decomposed. Remove plant debris or infected plant parts after each growing season. Turn the soil after harvest to help break down small roots that may harbour nematodes, fungi or bacteria. Gardeners may compost dead plants if they have a good composting system; otherwise, these piles may serve as a source of pathogens.

Prune or remove twigs and branches of woody plants affected with fire blight and other bacterial or fungal canker diseases.

Keep gardens weed free. Weeds often are another source of pathogens. Eradicate weeds to break the life cycle of pathogens and control them. Weed removal also can increase air movement and thus decrease conditions that favour disease development.

So that pathogens do not spread from one area to another, always disinfest machinery and other tools with steam, hot water under pressure, or a 10 percent solution of household bleach diluted with water.

CROP ROTATION

Avoid planting the same crop in the same area of the garden year after year. Continuous culture of the same kind of crop provides an opportunity for pathogen buildup.

For example, rotate leafy vegetables with grains or corn, or rotate annuals or biennials in seed and flower beds. It is best to grow the same or closely related plants in the same soil only once every three to five years. This practice starves out most pathogens that cause leaf, flower and stem diseases.

Crop rotation is not as effective against soilborne organisms, those fungi, bacteria and nematodes that persist in the soil for up to 10 years or more.

SOIL SANITATION TREATMENTS

Occasionally, disease-causing organisms that live in the soil may build up and prevent satisfactory growth of plants. Pathogen-free soil is desirable for houseplants, transplants and garden plots. Sterile potting mixes are available at many garden centres. However, it may be desirable to sanitize small quantities of soil on your own.

There are several nonchemical methods available to eradicate or reduce pathogens in the soil. The use of dry, steam or solar heat are the most effective nonchemical means to sanitize soil. The time to treat soil is before seeding or transplanting. Soil to be treated must be easily crumbled and without clods or large pieces of plant debris. Add any amendments (manure, compost, peat moss, etc.) before treatment. Soil also must have proper moisture. To test for this, gently squeeze a handful of soil. When the hand is opened, the soil ball should break apart somewhat. If it doesn't and the ball cannot be broken apart by gently pushing down on top of the ball, the soil is too wet. Oven sterilization. Place soil evenly but not more than 4 inches deep in nonplastic containers, such as seed flats, clay pots, and glass or metal baking pans. Tightly cover each container with aluminium foil. Insert a meat or candy thermometer through the foil into the centre of the soil. Set the oven temperature between 180 and 200 degrees F.

Heat the soil to at least 180 degrees and allow it to remain at this temperature for 30 minutes. Do not allow the temperature to go above 200 degrees, because this may cause products that are toxic to plants. After heating, cool and remove containers from the oven. Leave the aluminium foil in place until you are ready to use the soil. The heated soil will give off an odor. Microwaves or outdoor cookers also can be used.

Pressure-cooker sterilization. Pour several cups of water into the cooker. Place no more than 4 inches of soil in shallow containers on a rack out of the water. Level the soil, but do not pack it down. Cover each container with aluminium foil. Stack the containers to allow steam circulation. Close the lid, but leave the steam valve open until all the air is forced out and steam begins to escape. Then close the steam valve and heat at 10 pounds pressure for 15 minutes. Turn off the heat, allow the containers to cool, and remove. Leave the aluminium foil in place until you are ready to use the soil.

Steam sterilization without pressure. Pour about 1 inch of water into the sterilizing container. Follow the soil preparation procedures listed earlier. Place filled soil containers on a rack that will hold them out of the water. Close the lid and bring the water to a boil. Open the lid just enough to prevent pressure from building up. When the steam begins to escape, continue boiling for 30 minutes. Turn off the heat and replace the lid. Remove the soil when cool.

Soil solarization. This method uses the sun's energy to heat small areas of soil to temperatures that are lethal to many soilborne organisms and weed seeds. Treat during the summer when there are high air temperatures and intense solar radiation. With a tiller or a shovel, loosen the soil to be treated, wet it and cover it with a thin, clear, polyethylene (plastic) film. Seal the edges of the plastic sheet with soil to prevent heat loss and retain moisture. Leave the plastic in place for several weeks. The longer the soil is exposed to the heat generated by the solarization process, the greater the kill of undesirable organisms. Avoid toxicity from heated soil. With heavier soils and soils that contain a large amount of organic matter (manure, compost or peat moss), a toxic effect from heat sanitation may occur. This can cause poor seed germination, plant growth abnormalities or plant death. The toxicity is caused by an accumulation of ammonium compounds, soluble organic compounds, minerals or salts when the soil is heated too long or at too high a temperature. If soil toxicity is a problem, heavy irrigation of the treated soil may leach out many of these substances. Storing the soil two to three weeks without a cover also reduces soil toxicity.

Cultural Management

Cultural management involves avoiding the onset of disease. To accomplish this without pesticides, create an environment unfavourable to pathogens.

- Don't work in the garden when plants and soil are wet. Spores and cells of disease-causing organisms can spread from one plant to another and initiate new disease. Wet soils are easily compacted, which can decrease the amount of oxygen in the soil.
- Make sure plants are spaced properly. Air movement decreases when plants are grown too close together. This allows moisture to remain on leaves for longer periods of time. Wider spacing in beds and landscape plantings promotes rapid drying after wet periods and stops development of foliage, flower and fruit pathogens.
- Avoid excessive soil moisture. Overwatering enhances seed decay, damping-off and root rot diseases. Try not to plant in areas that have poor drainage or where water stands for several days following rains.
- Fertilize plants properly based on soil nutrient analyses using either organic or commercially prepared (inorganic) fertilizers.

In other words, use good cultural management. Healthy plants are less likely to have disease problems than weak, undernourished ones. Grow plants under optimum conditions and there will be fewer disease problems. Control of most plant diseases can be accomplished without pesticides. Use sound cultural practices, sanitation and well-adapted plant varieties to reduce disease problems. It is important to realize that you must accept some disease loss. Don't expect a perfect garden or plant if you do not want to use chemicals.

DISEASE CONTROL IN THE HOME VEGETABLE GARDEN

Plant diseases can be a significant problem in home gardens. Most vegetables are susceptible to a number of diseases. Wilts, leaf spots, blights and fruit rots are just a few of the disease problems that plague vegetables gardens every year.

Plant diseases are caused by four primary types of organisms: fungi, bacteria, nematodes and viruses. These organisms are often referred to as *plant parasites* or *pathogens*. Plant diseases caused by fungi and bacteria are most common when rain showers and/or heavy dews are frequent and temperatures are warm. During these times, scout the garden regularly for disease. Viral diseases can occur at any time, and nematode damage can be more evident if conditions are dry. Home gardeners can reduce the occurrence of many diseases with sound cultural practices.

Site Selection

Although few people have an ideal garden spot, site selection is important for a successful garden. A home garden site should be well drained. Avoid wet and poorly drained soils. Excessive soil moisture will favour root and crown diseases such as damping-off, crown and root rots caused by fungal soilborne pathogens. Most vegetables require full sunlight for maximum yield. Full sun speeds drying of the foliage, which reduces the incidence of most foliar diseases.

Crop Rotation

Crop rotation is very important in reducing losses to vegetable diseases. Continuous plantings within the same plant family of vegetables provide opportunities for pathogen buildup. Do not plant the same family of vegetable in the same areas year after year. Grow the same plants or closely related plants in the same soil only once every 3 to 5 years. This practice starves out most pathogens that cause stem and leaf diseases.

Unfortunately, crop rotation does not work against soilborne problems such as root and crown diseases caused by the fungi *Phytophthora, Rhizoctonia, Pythium* and *Sclerotium,* and vascular wilts caused by the bacteria *Pseudomonas*. These organisms are long-lived in the soil and affect many plant families. You can get only a limited benefit against *Fusarium* wilt.

DISEASE FREE SEED AND TRANSPLANTS

Many plant diseases can be seed-borne. Do not save seeds from year to year. This is important to prevent a number of diseases, including halo blight, common blight, anthracnose of snap beans, and bacterial spot of tomatoes. Commercially available seed is often produced in the arid western United States where seed-borne pathogens do not occur and/or cannot survive.

Seed of some vegetable varieties, however, are not produced in the west, and local seed must be used. Purchase seed from a reputable dealer because you cannot distinguish healthy seed from diseased seed.

Seed are protected by fungicides often applied to the seed by the company. Seed can also be treated by the home gardener.

These treatments give young seedlings some protection from soilborne pathogens as they germinate and emerge. They do not, however, control pathogens that may attack the plant after the seedling stage. Some pathogens are carried on the surface of seed and can be controlled with fungicide treatment. Fungicides are not effective on pathogens carried beneath the seed coat. Gardeners starting a crop from transplants should examine transplants carefully before purchasing. Select healthy transplants that appear disease and insect free.

Use Resistant Varieties

Using resistant varieties is the most efficient way of controlling vegetable diseases. Make an effort to buy resistant varieties when they are available. Seed catalogues generally list resistant traits of different vegetable varieties. For example, disease resistant tomatoes are designated "VFN." These letter indicate resistance to Verticillium wilt (not common in Georgia), *Fusarium* wilt and nematodes.

In tomatoes, there is also resistance to *Stemphyllium* (gray leaf spot) = S, Tobacco Mosaic Virus = TMV, *Alternaria alternata* stem canker = A or ASC, *Septoria* leaf spot = L and Tomato Spotted Wilt = TSW.

Other vegetables such as squash, pepper, beans and peas also have disease and nematode resistance. Note that plant disease and nematode resistance are relative terms, as resistant varieties have differing levels of resistance. Resistance does not always mean the plants are immune to these problems. Some seed catalogues may indicate tolerance to a certain pest. Tolerance usually indicates that a particular variety of vegetable may produce fruit of acceptable yield and quality while exhibiting moderate to high levels of disease.

PLANTING DATE MANAGEMENT

Planting dates can be an effective tool for disease management. Follow the recommended planting dates for the particular vegetable grown. For example, okra should be planted when soil temperatures are warm for good germination and growth. Planting this crop when soil temperatures are cool can cause increased incidence of soilborne diseases such as soreshin and damping-off. Using transplants may help avoid early disease problems so long as the transplants are from a reputable source and are disease free. Plant crops such as corn and beans earlier, because their seeds germinate in cooler soils. It is also advantageous to plant these particular crops early to escape severe virus infections. This is particularly true with squash. Aphids, which transmit many virus diseases, are at lower population levels early in the season. High aphid populations late in the season can lead to more virus problems.

Trap Crops

Using trap crops may sometimes help manage virus diseases and aphid populations. A few rows of a trap crop around the vegetable garden (such as rye or corn) will cause aphids to feed there first, possibly loosening the virus they may carry. Trap crops can limit aphid damage and help reduce the incidence of virus diseases, which are sometimes severe on beans, cucurbits and solanaceous crops.

Proper Spacing and Trellising

Proper spacing and trellising can reduce the occurrence of vegetable diseases. Space plants properly to allow growth and air circulation. Humid and wet conditions occur if plants are crowded and unable to dry quickly. Extended periods of wet foliage are ideal for development of fungal and bacterial diseases. Crowded plants take longer to dry and provide favourable disease conditions. Crop protectants also cannot penetrate dense foliage, thus limiting their effectiveness. Staking or trellising tomatoes, pole of half runner beans, and cucumbers will prevent soil contact with the foliage and fruit. Air circulation will be better if these plants are trellised, promoting foliage drying and reducing disease severity. Limiting soil-plant contact reduces diseases such as fruit rots. Crop protectants also can be more effectively applied to trellised plants.

Proper Watering Practices

Avoid excessive soil moisture. Over-watering enhances seed decay, damping-off, and root and crown rot diseases. *Choanephora* fruit rot, which causes the whisker-like blossom end rot, can be increased by over-watering. If possible, irrigate by running water between the rows. Avoid splashing plants with soil, since any potential pathogens are soilborne. Drip irrigation is also a good method of watering without wetting foliage. Drip irrigation uses either a hose or tape that allows water to be applied slowly over time at the base of the plant or just beneath the soil surface. Garden drip irrigation kits are readily available. Avoid wetting the foliage when watering. Wet foliage is favourable for development of most foliar fungal and bacterial diseases. If wetting the foliage is unavoidable, irrigate in the morning so foliage can completely dry before evening.

Maintain uniform soil moisture. This can reduce problems such as blossom end rot (calcium deficiency) of tomatoes and peppers. Don't work in the garden when plants and soil are wet. Bacterial and fungal diseases spread easily from one plant to another by hands and clothing when above-ground plant parts are wet.

Use a Mulch Layer

Use a mulch layer of straw, bark, shredded paper or plastic to prevent soil from splashing onto plants and to prevent fruit from touching the bare ground. This will help prevent rots on mature fruit such as strawberries, tomatoes, squash, cucumbers and melons. Mulches also are a sound cultural practice to help conserve soil moisture and reduce weed infestations.

Proper Fertilization

Adequate and proper fertilization helps prevent vegetable diseases. Test soil three to six months before the growing season, and follow the recommendations to supply appropriate nutrient requirements and adjust soil pH. Avoid excessive amounts of nitrate forms of nitrogen, which encourage root rot diseases. Use ammonium forms instead (no more than 50 percent). Be sure the soil pH level is in the proper range for a particular crop. Proper pH prevents blossom end rot and encourages healthy growth of tomatoes and peppers.

Weed Free Garden

Weeds in the same families as some of the vegetable crops can be another source of diseases and insects. For example, some weeds can serve as virus reservoirs for several insect-transmitted viruses that can infect homegrown vegetables. Good weed control will increase air movement in the garden and decrease conditions — such as excessive moisture — that favour disease development.

AVOID TOBACCO WHEN WORKING IN THE GARDEN

If you use tobacco, wash your hands thoroughly before handling plants. This practice will prevent the spread of tobacco mosaic virus, which can infect many different kinds of vegetables, particularly solanaceous crops such as tomatoes and peppers.

Nematode Control

Nematode control is important in growing vegetable crops. Nematodes are microscopic soil-inhabiting roundworms that injure vegetables and other plants by feeding on their root systems, causing decay and/or galling. You can sample soil for nematodes by submitting a sample through your county extension office to the Extension Nematology Laboratory. The most damaging nematode in Georgia is the root-knot nematode, named for the knots it produces on infested roots. Soil solarization can be used to reduce the number of nematodes present. This involves tilling the garden and then covering the area with a clear plastic tarp for six to eight weeks. Solar radiation is trapped in the soil and can raise soil temperature high enough to kill most nematodes in the upper 3 to 4 inches of soil. In Georgia, the best time to solarize soil is June through August, when temperatures are at their hottest. Soil solarization requires intense solar heat to be effective. Repeated roto-tilling can also desiccate and kill many nematodes, although repeated tilling may decrease overall soil organic matter.

Pesticide Use

Use of pesticides should be the last defence available to home gardeners once all other disease-control options have been exhausted. Home gardeners have few fungicides or bactericides available compared to commercial growers. Chlorothalonil (Bravo® or Daconil®), maneb, mancozeb, Terraclor® (PCNB), sulfur and copper products can be used on certain crops. These products provide disease suppression across a range of foliar diseases and are more effective if they are applied in a preventive manner at the very onset or before a disease outbreak occurs. Chlorothalonil, maneb and mancozeb generally provide broad-spectrum disease suppression of most fungi. Terraclor is used as a transplant soil drench to suppress Southern blight and Rhizoctonia damping-off. Copper products suppress fungi but primarily reduce losses to bacterial pathogens. Sulfur is suppressive to fungi and is especially effective against powdery mildews. As with any pesticide, read the label and follow the recommended precautions to ensure the chemical is applied in a safe, effective manner. Contact your county extension office for specific information about pesticides.

INFECTION BY PATHOGENS

Pathogens can infect plants in several ways Fungi and bacteria may penetrate (enter) through natural openings such as the stomata — these are tiny "breathing" pores in the leaf that allow the exchange of gases. Viruses, fungi, and bacteria also enter through wounds. Bacteria frequently enter through hail wounds. Many fungi can penetrate the leaf surface directly without wounds or stomata.

Some viruses and most phytoplasmas are carried by insects, especially sucking insects such as aphids and leafhoppers. Insects that carry and transmit plant disease organisms by their feeding are called vectors. The aster leafhopper, for example, is the vector of both the aster yellows phytoplasma and the oat blue dwarf virus.

Some viruses are transmitted mechanically by the rubbing together of leaves or by humans touching diseased and then healthy leaves. Tobacco mosaic virus and potato virus X are common examples. Many fruit tree viruses are transmitted by grafting.

Many pathogens, especially foliar (leaf) pathogens, need a film of water on the plant to begin growth, penetrate the host, and establish infection. This is why wet or humid weather is so important in the development of many fungal and bacterial diseases.

DISSEMINATION OF PATHOGENS

Pathogens are disseminated (spread) by wind, insects, water, man, animals and birds.

Wind

Wind disseminates fungus spores from plant to plant in a field or across fields. Pathogens such as the wheat leaf rust, wheat stem rust, and barley stem rust pathogens are spread long distances by the wind. Leaf rust is wind-borne from the major winter wheat areas of Kansas, Nebraska and Oklahoma (Occasionally it survives the winter in North or South Dakota). Stem rust is wind-borne from Mexico and Southern Plains states to wheat and barley crops in the Northern Great Plains.

Insects

Insects are important in carrying viruses and phytoplasmas from southern areas. Common insect vectors include aphids for the barley yellow dwarf virus and the aster leafhopper for the aster yellows phytoplasma. Insects also spread these pathogens from plant to plant. Bacterial wilt of cucumber and muskmelon is disseminated by both the striped and spotted cucumber beetles; only the striped cucumber beetle is common in North Dakota.

Water

Water can carry pathogens from field to field. Rain and splashing water can disseminate many fungi and bacteria. Septoria leafspot of tomato and the bacterial blights of dry beans are common examples. Rain and wind form numerous tiny airborne water droplets called aerosols. Many bacteria are disseminated long distances in wind-driven aerosols. Water flowing over the surface of fields spreads disease organisms such as Sclerotinia (white mold), *Verticillium,* and downy mildew.

Man

Man spreads pathogens and weeds over long distances. International plant quarantines attempt to prevent this dissemination by authorizing inspection of planes, ships, cars, and luggage for prohibited pests at ports of entry. The Mediterranean fruit fly as well as many weeds and pathogens are frequently intercepted at ports of entry.

The gardener who brings fruit in his or her luggage and the commercial grower who brings seed of a high performance crop in his pocket may introduce a new pathogen or other pest. Man can also disseminate pathogens and other pests locally. Dutch elm disease was originally introduced into the United States on elm logs; much of its local spread and buildup has been on elm firewood collected from diseased trees. The soybean cyst nematode, not yet reported (2000) in North Dakota, could be introduced on contaminated farm implements from infested areas in neighboring states. Man can locally disseminate bacterial blights, rust and anthracnose of dry beans by cultivating the crop when it is wet.

Animals and Birds

Animals and birds also may disseminate pathogens. For example, the soybean cyst nematode can be disseminated in the feces of birds; animals may spread water-borne pathogens by walking through an infected crop when the plants are wet.

SURVIVAL OF PATHOGENS

In North Dakota we are primarily concerned about pathogen survival between crop seasons. When this is over the winter, it is called overwintering.

Soil

Many pathogens form resistant structures that survive long periods of time in the soil.

For example, the Aphanomyces root rot pathogen of sugarbeets survives for over 20 years as resistant spores in the soil, and the sunflower downy mildew pathogen survives for at least 14 years as resistant spores in the soil.

The white mold pathogen of dry beans and sunflower survives for six to eight years or more as resistant bodies in the soil. The Verticillium wilt pathogen of potatoes, tomatoes and other crops survives for at least several years in the soil.

Plant Parts

Pathogens may survive on crop refuse. These include the leafspot pathogens of wheat and barley which survive on stubble, and the tomato Septoria leafspot pathogen which survives on the old dead vines. Destroying or burying this crop refuse reduces next year's disease potential.

Seed and Vegetative Plant Parts

The smuts of small grains survive on or in the seed and survive in storage as long as the seeds remain viable (can germinate). Vegetative plant parts such as tubers, roots and corms provide a mode of survival for many pathogens. Many potato pathogens are carried on or within the tubers.

Insects and Mites

The bacterial wilt pathogen of cucumber and muskmelon is suspected to overwinter in the digestive tract of cucumber beetles. The wheat streak mosaic virus overinters in wheat curl mites that survive on winter wheat and some perennial grasses.

Mild Climates

The cereal rust pathogens usually do not overwinter in North Dakota but survive year-round in the southern United States and Mexico and are carried north each year by wind.

HOW DISEASES ARE MANAGED

This section focuses on management of infectious diseases. Non-infectious diseases are not discussed as their management involves a remedy of the physical factors that induce them.

The four basic methods of infectious disease management are: exclusion, eradication, host resistance, and protection (including the use of fungicides). These four methods reduce pathogen populations or slow their development. Protectant fungicides and resistant varieties slow down the development of pathogen populations. Cultural practices also can reduce the pathogen's population. In short, management practices prevent or delay the introduction of pathogens or reduce initial pathogen populations and retard their subsequent increase.

Sound management is based on correct diagnosis. This is essential to distinguish infectious from similar appearing non-infectious diseases as well as to correctly identify the pathogen involved in infectious disease. Correct identification of the pathogen is essential to know the pathogen's life cycle and how it relates to the cycle of disease development. This information is needed to develop a management program that attacks the pathogen at the weakest point in its life cycle. When fungicides are used, the type and the timing are important. For example, in the case of dry bean diseases three different types of fungicides are used for management of each of three major diseases: rust, bacterial blights, and white mold. In each case timing is important and application must be started before the disease is widespread. For head scab of wheat and barley, timing of fungicide application is critical as well, at early heading for barley and early flowering for wheat.

EXCLUSION

Exclusion means exclusion of pathogens. Pathogens can be excluded (or kept away) from hosts by quarantines that prevent their introduction, and by use of seed stocks certified to be pathogen-free or within certain prescribed tolerances for low levels of pathogens.

QUARANTINE

International quarantines are familiar to anyone who has travelled overseas. Planes, cars, trucks and luggage are checked at ports of entry to prevent the introduction of pathogens and other pests into areas where they do not occur. Many disease organisms would flourish in our state if they were introduced. State and local quarantines are used to keep black wart of potato and the golden nematode restricted to a few localized areas of the eastern United States. International quarantine has prevented new introductions of pathogens from other countries.

SEED CERTIFICATION

Seed certification is used to certify that potato seed tubers and seed of dry beans have low levels of pathogens or are pathogen-free in the case of potato ring rot. The crop is grown from seed produced under carefully controlled conditions. Sometimes the seed crop is grown in an isolated area to reduce disease potential. For example, foundation seed potatoes are grown in Golden Valley County, a western county isolated from the rest of the potato production in North Dakota. Crops are field inspected and must meet certain tolerances to be certified. Certification may be done in conjunction with indexing. Many seed potatoes are now produced in greenhouses.

INDEXING

Indexing involves laboratory or greenhouse tests to determine infection by pathogens in vegetatively propagated plants such as potatoes and fruit trees. Only the healthy materials are saved for further increase.

Culturing

Plant parts to be used for increase are laboratory cultured to determine if they are infected with pathogenic fungi or bacteria. Plant parts found free of pathogens are used for further vegetative increase and propagation. This technique is used on chrysanthemums, carnations and potatoes.

Indicator Hosts

Indicator hosts are plants that produce rapid and distinctive symptoms when inoculated with a virus. They are used to detect specific viruses in individual plants. Fruit trees and potatoes are commonly indexed for viruses. Only those trees free of virus are used as sources of budwood for graft propagation. Indexed nursery stock is commonly used for orchard plantings, as this is the only practical way to control fruit tree virus disease.

Serology

Serology is the use of specific antibodies present in the antiserum of warm-blooded mammals. These antibodies are produced in the blood of rabbits (usually) that have been immunized with a specific pathogen. This procedure is used to index barley and potato seed stocks for specific viruses.

It is also used to detect some bacteria. Many highly specific and sophisticated tests are available now.

Embryo Test for Barley Loose Smut Detection

The embryo test determines if loose smut is present and, if it is, the percentage of infection. If the loose smut infection exceeds 1 or 2 percent the seed lot should be treated with an effective systemic fungicide prior to planting. The embryo test is not available for use on wheat seed.

ERADICATION

Eradication means elimination of the pathogen. In actual practice, this term may be used when the pathogen is not completely eliminated but the populations are greatly reduced.

CROP ROTATION

Crop rotation involves growing different crops in the same field or plot in succeeding years. Pathogens such as the fungi that cause tan spot of wheat, dry bean rust, Cercospora leafspot of sugarbeet, and Septoria leafspot of tomato attack only one host, and populations of the pathogen increase when the same host is grown repeatedly on the same land. Crop rotation helps keep populations of these pathogens at low levels. There is one precaution, however: the grower must consider nearby areas as well. Disease organisms may spread from nearby fields or garden plots if disease was present in those areas the previous year. Disease can be expected to develop first in the area next to last year's crop. If weather favours disease development, the entire field or garden plot may eventually become diseased.

Crop rotation is an effective tool for reducing many pathogen populations. However, some pathogens survive many years in the soil and are not affected much by normal crop rotations. Long rotations may be necessary but often are impractical. The sunflower downy mildew pathogen, the sugarbeet *Aphanomyces* pathogen and the white mold organism survive many years in the soil. The pathogen that causes Verticillium wilt of tomato survives several years in the soil and also attacks many other garden vegetables, so it is difficult to eliminate by rotation. In the case of Verticillium wilt in tomato, use of resistant host varieties is the practical solution. In the case of white mold, some navy, kidney and black beans show partial resistance to white mold. In the case of sunflower downy mildew, only a few hybrids are resistant to all races of the mildew fungus.

ERADICATE ALTERNATE HOSTS

Many rust fungi require two hosts to complete their life cycle. The second host, called the alternate host, is essential to overwintering of many rusts in northern climates. Some of these same rust fungi reproduce indefinitely without the alternate host in warm climates. Examples of alternate hosts include common barberry (not ornamental barberry) for wheat and barley stem rust, buckthorn for oat crown rust, and juniper for apple rust. The rust fungi's sexual phase occurs on these alternate hosts.

It was hoped that eradication of barberry from the Upper Midwest in the 1930s would break the pathogen cycle and eliminate the stem rust fungus. After most barberry had been eradicated, stem rust still occurred. It was determined that stem rust survives year-round in Mexio and the Gulf Coast in the summer spore stage, without requiring the barberry for the overwintering stage. These spores are wind-blown thousands of miles north to the Upper Midwest every year. Nevertheless, the barberry eradication program had two very important accomplishments: 1) stem rust infections started later in the season, and 2) the sexual phase of the fungus was eliminated, which slowed down the development of new rust races. Elimination of buckthorn near oat fields produces similar results. To manage apple rust, commercial growers try to remove all junipers within two miles of their orchards. This is not feasible for the homeowner, who must use fungicides or resistant varieties for management of apple rust when weather is wet in spring and early summer.

SANITATION

Sanitation is the removal of crop refuse. Tillage is sometimes used to bury the refuse. The quantity of a pathogen available to produce infection is called the inoculum. Burial by tillage reduces the inoculum of the wheat tan spot pathogen, the barley spot blotch pathogen, and many garden pathogens.

All diseased tomato vines and refuse should be removed, and apple leaves that had apple scab should be raked up and detroyed by burning, burying or sending to the landfill.

When Septoria leafspot of tomato is severe, picking off badly diseased leaves (sanitation) before spraying with a fungicide helps reduce the inoculum and improves fungicidal control.

HOST RESISTANCE

Resistance is the ability of a host to resist infection by a pathogen. Resistant varieties are favoured by commercial growers and gardeners when they are available. Resistance has been the best and most cost-effective method of managing stem rust and leaf rust of wheat. Homeowners who have the Verticillium wilt pathogen in their garden soil must grow a resistant tomato variety to manage the disease.

Some foliar (leaf) pathogens may rapidly develop new races quite capable of attacking certain types of host resistance. Many foliar pathogens are extremely variable and produce billions of spores that are disseminated great distances by the wind.

Consequently, new races of some foliar pathogens may become widespread in a short period of time. This results in disease outbreaks and a continued need for plant breeding programs.

Soil-borne pathogens (root and vascular wilt pathogens such as Verticillium) are also variable, but new races may not become widespread as quickly. There are two types of host plant resistance: race specific resistance and general resistance.

Race specific resistance usually provides a high level of resistance, but it fails when new races of the pathogen develop. In the late 1990s new races of wheat leaf rust developed that attacked some of the previously resistant wheat cultivars.

General resistance is usually a stable type of resistance that is effective against all races of the pathogen. General resistance usually does not exhibit as high a level of resistance as race specific resistance. However, general resistance slows down disease development compared to that on a susceptible variety.

PROTECTION

Protection means protecting plants from infection. Storing potatoes and other vegetables in cold storage protects against infection because it is too cold for many pathogens to develop, or development is greatly slowed down. Seed potatoes are grown in isolated areas where aphid populations are low (exclusion) and thus easily managed (protection); this minimizes aphid-borne virus infection.

CULTURAL PRACTICES

Time of planting may help plants escape infection. Winter wheat is planted in September after the destruction of volunteer wheat. Destruction of volunteers prior to winter wheat planting destroys the green bridge that wheat curl mites survive on between the summer and fall crops. The wheat curl mite is the vector of the wheat streak mosaic virus. Delayed planting of winter wheat also reduces the risk of survival and buildup of the mite in the fall of the year and exposure of the wheat crop to high wheat curl mite populations. Many dry bean and garden bean pathogens are disseminated in water, so beans should not be cultivated when they are wet. A plastic mulch used on tomatoes greatly reduces blossom end rot, a non-infectious disease that develops under conditions of drought or fluctuating soil moistures. The mulch produces a more uniform soil moisture. Most powdery mildews are favoured by high humidity. Powdery mildew is a common problem on alpine current, lilac, roses and shaded lawns. Pruning shrubs and trees to allow better air circulation and sunlight penetration may help reduce powdery mildew in shady locations.

HANDLING PRACTICES

Development of potato late blight in storage can be minimized by proper handling practices during the growing season. This includes hilling the soil around the plants to reduce the chances of late blight spores coming into contact with the tubers. The vines should be killed several weeks before harvest by using approved vine killers or chopping the vines off at ground level. The late blight fungus on the tops will be minimal at harvest and tubers will be mature. Mature tubers are less prone to infection.

MANAGING INSECT VECTORS

Many insects carry disease organisms such as viruses and bacteria. Managing these insect vectors may reduce the chance of disease. A prime example is the cucumber beetle, vector of the bacterial wilt pathogen of cucumber and muskmelon. The bacterium is carried from plant to plant by the beetle and overwinters in the beetle. A good program for managing the cucumber beetle, started as soon as the plants emerge, will prevent serious losses from the bacterial wilt disease. Similarly, good aphid management is essential for raising virus-free seed potatoes. In addition to successful management of insect vectors, weeds and other hosts that can serve as a reservoir for both vectors and pathogens must be managed. Weed management is essential around plantings of potatoes, tomatoes, cucumbers, melons, peppers and many other commercial and garden crops.

4

Diseases Caused by Fungi

In keeping with a general tendency in plant pathology, most of the parasites able to attack the genus Populus are fungi, primarily Ascomycetes, but also many Mitosporic Fungi and Basidiomycetes

The number of fungi reported at least once on diseased poplars is so vast that to cite each one would itself be a chimerical task and in any event outside the scope of this work; new host-parasite associations, however, are being constantly described. Attention will thus be directed to the agents of the most important diseases on account of their incidence and geographical distribution, as well as of those known to be endemic in a given region or typical of certain soils and climates, and of those whose range is likely to extend in the future.

Generally speaking, the conventional distinction between root diseases, stem and branches diseases and leaf diseases is still useful, even though some parasites (such as *Septoria musiva*) are able to infect more than one organ, while leaf diseases, especially in intensive growing, result in poor trunk growth and hence in diminished production of wood. Even if it is not actually killed (an extreme event that occurs above all in the nursery or in the first 1 or 2 years after planting out), the whole of a tree suffers the consequences of infections of one of its organs, and it is also rendered more vulnerable to environmental stresses.

The fungi that cause leaf diseases are almost always primary parasites, i.e. able to attack healthy plants, whereas those that attack trunks and roots are mostly wound parasites, thus requiring breaches in the bark to be able to infect, and/or weakness parasites, which means that they are only aggressive on plants already debilitated for some other reason. Depending on the circumstances, therefore, they can take advantage of small more or less accidental injuries caused during the various cultural practices, or of a poor management, or of the unsuitability of the soil and/or bad climatic course. Their attacks may also follow previous infections of primary parasites or other weakness parasites on the same hosts. Therefore, a combination or succession of diseases is by no means uncommon in precarious and simplified ecosystems such as intensively cultivated poplar stands.

Although they are quite similar as regards their type of parasitism, root pathogens and stem pathogens give rise to different forms of damage: the former, through a physiologic impairment of the infected plants due to the rots of absorbing tissues, cause predominantly quantitative wood losses, the latter also result in a marked loss of quality, through the chromatic and/or structural alterations of the wood rings below the bark tissues affected by necroses or cankers.

When compared with many more strictly forest genera, the large number of epidemic diseases to which Populus is subject may come as a surprise. Nevertheless poplar growing, whether in the nursery or in the plantation, involves the establishment of ecosystems that are extremely favourable to the rise of epidemics, either because plants are crowded together and often lack genetic diversity on vast areas due to the preferential employment of some clones, and because the array of fungous competitors that, in natural stands, help to maintain many diseases in a latent form here is very rarefied.

Lastly, it is interesting to notice that most pathogenic fungi on poplar display a certain specificity for section Leuce Duby, or for sections Aigeiros Duby and Tacamahaca Spach together, in keeping with their high affinity (they only differ in some vegetative characters). Little information is as yet available concerning the diseases of poplars belonging to sections Turanga Bunge, Leucoides Spach and Abaso Ecken.

DISEASES OF STEMS AND BRANCHES

Bark Necrosis Caused by Discosporium Populeum

Together with Septoria musiva, Discosporium populeum, also known as "Dothichiza", is one of the most widespread weakness parasites affecting poplars of section Aigeiros used for intensive cultivation. It is not improper to say that the measures adopted to combat its attacks, improving agronomic techniques, have increased the production through a rationalisation of some of the cultivation stages.

Symptoms and the damage caused-Necroses caused by D. populeum can impair the growth and often the survival of hosts during the whole vegetative season after the infections, though they usually begin in late autumn, or in late winter in the case of one-year saplings, when the same hosts are already in the resting stage. The target is one-and two-year-old nurseries or plantations up to three-years-old subjected to water stresses, often because they are located on too sandy soils prone to periodical water shortages, or due to the onset of a transplantation crisis or the concomitance of these two causes. At all events, a predisposing condition to the progression of infections is a dehydration of the cortical tissues that will subsequently be aggravated by the parasite.

The first visible symptom is the appearance of variously sized patches, pale maroon at first, then turning to brown, on parts of the bark near its various discontinuities: leaf scars, insertion points of buds and branches, separation rings between the annual shoots, abrasions, injuries or prunings. These patches, symptomatic of incipient necroses of large areas of the bark, soon extend, mainly along the axis of the infected organ, and their surface often sinks below the level of the surrounding healthy tissues due to progressive dehydration of the outermost layers. Exposure of the bark inner layers and of the adjacent wood circles shows that they are brown and marcescent, and sometimes damper than the non-infected areas.

At a later stage of the disease, some sections of the dead bark are lifted up by the internal pressure of the imperfect fruiting bodies that have formed in the meantime and soon split above them and expose them to atmospheric agents. These fruiting bodies, known as pycnidial conidiomata, are by now visible to the naked eye as coriaceous, blackish and globose corpuscles (Ø = 1-2 mm) with a small opening for dispersion of their conidia, erupting from the periderm of the stem or branch attacked, and often arranged in rows or in concentric circles around the penetration point of the parasite. If the weather is sufficiently humid, the conidia are released in a yellowish, mucilaginous substance and dispersed by the rain and the wind.

When the host starts to grow again, it can circumscribe the infection and react against it by forming callus tissues, if the water capacity of the soil has returned to optimum levels and if the portions of damaged tissue are not too many and/or too extensive, though complete healing is rather rare. In this stage, the disease takes on the appearance of a "canker" that in any event points to an active role on the part of the plant.

On other occasions, especially if the state of hydro-physiological imbalance persists and the host does not have the resources needed to mount a reaction, the parasite kills other areas of the bark during the spring. These killed areas eventually occupy the entire circumference of the trunk and lead to the death of the part of the plant above. In the most serious cases, the annular extension of the necrosis is completed when the plant is still dormant and even the spring renewal of growth is lost.

Heavy attacks by D. populeum can both kill off much of a nursery's stock and also impair the further growth of the plants that remain and their subsequent planting out. Furthermore, plants that get over the critical stage of the infection run the risk of being broken off by the wind at their colonised areas. The quantitative damage is thus translated into loss of productivity or fewer poplar saplings available for sale. Nevertheless, adult plants with extensive infections have been observed in Italy in recent years, mostly in portions already showing the so-called "brown spots", reported on the "Luisa Avanzo" Euramerican clone, which is known to be sensitive to hydro-physiological imbalances. Since the brownings linked to the necroses in this case are peripheral when the wood is used, there is a loss of quality and hence a substantial decrease in the price of the ply-wood products.

The pathogen-Discosporium populeum (Sacc.) Sutton [= Chondroplea populea (Sacc.) Kleb. = Dothichiza populea Sacc. et Briard] is the name given to the anamorph of Cryptodiaporthe populea (Sacc.) Butin (fam. Valsaceae, ord. Diaporthales, phylum Ascomycota), whose fruiting bodies, once rarely found on dead organs, are observed with greater frequency today.

The parasite is found in most parts of the world. Its areale, whose extension has certainly been favoured by human activities, embraces Eurasia (India and China included), North Africa, North America and Argentina. There have been several epidemics, both in south-central Europe during poplar cultivation dawning at the start of the 20th century, and after the Second World War in most of Central Europe, particularly Bulgaria, Poland and the former Jugoslavia. Recrudescences in some poplar regions (e.g. in northern Italy) in the 1980s were attributable to the drier climate and the dissemination of new Euramerican clones, that were introduced on account of their resistance to Marssonina brunnea and their appreciable technological characteristics, but proved to be susceptible to D. populeum owing to their insufficient environmental adaptability.

The virulence of D. populeum seems confined essentially to Aigeiros section. It attacks both P. × euramericana, to an extent that varies from one clone to another, and P. nigra, whereas P. deltoides is more resistant (indirectly, due to its greater tolerance of water shortages) as are P. trichocarpa and the P. deltoides × P. trichocarpa and P. deltoides × P. maximowiczii hybrids.

Biology and relations with the host-Hydro-physiological imbalances, due to a substantial lack of rain and/or a too coarse texture of soil and/or the excessive competition for light and nutrients, frequent in nurseries or quick-turnover plantations, are predisposing to D. populeum infections of the host because of the more or less rapid dehydration of its cortical tissues. This is a necessary, though not sufficient prerequisite for the effective triggering of an attack: for this to happen, in fact, there must be penetration through a discontinuity of any kind in the bark, whether natural or artificial (referred to above when speaking about the location of the symptoms), that exposes the underlying vital tissues. Contamination usually takes place in spring; however, autumn is the best season for inoculation and infection, which proceeds during the following year. There are some reasons for this prevalence of autumn infections:

- the ability of the plant to react is reduced, especially as the result of the long shortage of water during the summer;
- the fungus can grow at low temperatures, and also takes advantage of the increase in relative humidity for the germination of conidia;
- germination itself is boosted by the long autumn nights.

Several studies have shown that temperatures of about 10 °C, common in this period, do not hinder the fungus; in addition, they inhibit proliferation of the host's secondary meristems so that it cannot circumscribe the colonisation of its tissues, which increases in vigour as dehydration becomes more pronounced.

Conidiomata typical of the anamorph (whose appearance, described above, is a useful diagnostic factor) develop in the dead bark in late winter or early spring. The one-celled conidia, hyaline and ovoid (10-13 × 7-10 μm), form a sufficient vital mass of inoculum during most of the year, since they are released gradually on several occasions and maintain their viability for a long time (estimated at about five years in the laboratory). As already mentioned, however, the parasite transmission take place in spring, when transfer to new hosts is facilitated by the abundant equinoctial rains and the wind.

The teleomorph is provided by black, spheroid (Ø = 500-600 μm) and long-necked perithecial ascomata, later differentiated from the conidiomata, within which hyaline, two-celled, ovoid ascospores (16-23 × 6-9 μm) are formed. Its recent greater frequency has been seen as evidence of an insurgence and diffusion of biotypes endowed with greater virulence than the old populations, as would seem to be confirmed in isolates from various poplar-growing areas of south-central Europe. These «hypervirulent» strains, which can be regarded as one of the reasons for the recrudescence of attacks along with the resort to predisposed poplar clones and with prolonged dry periods, are also a cause of further concern for future developments of the disease, since it is well known that sexually-based genetic recombination is a potential source of phytopathological dynamism, with practical consequences readily imaginable.

Control strategies-It is necessary to exploit the resistance to D. populeum displayed by some North-American and Asiatic poplar species, as well as by intersectional hybrids. At the same time, as much as possible must be done to eliminate some of their basic adverse features, such as the fragility of the root apparatus and the vulnerability to the wind of P. deltoides, or the susceptibility to Melampsora medusae of some genotypes of P. trichocarpa or P. deltoides × P. maximowiczii. This objective can be pursued through a rational genetic improvement, such as the crossing of resistant P. nigra genotypes, which yet exist, with the aforesaid species to obtain clones free from the defects typical of their parents. Account must none the less be taken of the fact that real resistance to bark necroses seems connected to a form of tolerance indirectly established by environmental adaptability of clones, rather than to the genetic stock.

While the use of plants with a constitutional resistance to the disease is an obvious step towards its prevention, equal attention should be directed to the application of rational cultivation practices ensuring that they are maintained in a good hydro-physiological state, especially in predisposing climatic or edaphic contexts.

The greater managerial and financial effort associated with the constant and careful observation of these practices will probably be offset by the positive outcome of cultivation and hence of production in the nursery, while defective or negligent management, though less burdensome, could result in the loss of an entire crop. In this connection, it must be made clear that curative measures fit to restore a normal hydration of stricken plants are little more than palliative, especially in a nursery, since the reaction of saplings is uncertain and they will be hard to market in any event due to the marks left by the callus tissues. In a plantation, on the other hand, phytosanitary pruning of infected branches to reduce the inoculum centres may be advantageous, provided the saplings are not exposed to further water shortage stress.

A correct prevention during cultivation should be sought through the following measures:

- establishment of nurseries on soils with good texture and structure to ensure sufficient water retention;
- adequate spacing of the cuttings to reduce the effects of reciprocal competition;
- irrigation, working of the soil, moderate use of fertilisers and of herbicides to prevent competition from weeds;
- prevention of leaf diseases (especially rusts) that weaken the host and predispose it to infections by D. populeum;

- uprooting of the saplings when it is certain that they are in the dormant stage;
- their speedy and careful transport, preventing accidental bark damage;
- their preventive immersion in water for 5-6 days, sufficient to re-establish optimum hydration;
- gentle planting out at the right depth (2-3 m in sandy soils and 1-1.5 m in clayey or loamy soils) and encouragement of early resumption of growth, e.g. by an application of nitrate fertilisers.

Chemical control may be required in the nursery. For cuttings from clones that grow vigorously, often more vulnerable to water shortage stress and hence predisposed to attacks of the parasite, it is advisable to integrate the aforesaid cultural measures with some preventive fungicide treatments, at first when a nursery is established and then twice a year between May and June. Solutions containing hexaconazole (3 mL/hL a.i.), carbendazim (25 g/hL a.i.) or chlorthalonyl (120 mL/hL a.i.) have proved effective for this purpose.

Biological control has not yet shown signs of producing satisfactory results.

Necroses and Cankers Caused by Cytospora Spp.

Like the better known Discosporium populeum, the Cytosporae are virtually cosmopolitan parasites and appear in stands subjected to some form or stress or poor agronomic management, often in the train of an attack by the former. Correct application of the culture measures against the same D. populeum, however, should also serve to extinguish the possibility of their appearance.

Symptoms and the damage caused-Infection begins in the late autumn or in winter, when the host is dormant. It strikes nurseries already stressed for other reasons, saplings prior to being planted out and recently established plantations. Adult trees too can be attacked by Cytosporae, if they are in a state of advanced decline, and in this case the fungus behaves as a distinctly secondary parasite.

The main species in terms of diffusion and severity of attacks is C. chrysosperma. It causes a slight depression and brown to blackish staining of the bark infected areas and of the underlying tissues, both symptomatic of an extending necrosis against which opposition on the part of the host is weak until the vegetative season begins. This is particularly true with regard to the more delicate stems and branches, which may succumb to withering if ringed by this necrosis, whereas the sturdier ones are often in time to build up a resistance to further invasion. The disease than appears as small brown depressions bounded by distinct calluses. On P. tremuloides and P.alba the fungus turns the bark orange beside the infection sites.

At the end of the winter, pycnidial conidiomata differentiated by the parasite in the dead tissues gradually push up the outer layers of the bark. These fruiting bodies are of diagnostic value, since they are clearly smaller (Ø = 0.5-1 mm) even to the naked eye than those of Discosporium populeum and more randomly arranged.

They remain within the cortical tissues of the host and only their ends, each with a minute orifice called an ostiole, are exposed to the environment, where they release masses of conidia suspended in linear or spiral mucilaginous cirri, orange-yellow in the case of C. chrysosperma, that are dissolved and dispersed by the rain.

In the advanced stage of the disease, the dead bark tissues that have not formed a callus peel away to reveal the underlying wood, marked with red-brown bands that eventually turn black.

Attacks by C. nivea and C. ambiens are usually more occult. Since there is no browning of the bark infected areas as in the case of C. chrysosperma, signs of infection are only visible when the conidiomata partly emerge on the non-darkened surface. Another diagnostic character is provided by the different colours of the mucilaginous cirri released by the three species

Except in some restricted geographic regions, the incidence of the Cytosporae is less than that of D. populeum. The kind of damage they cause is exactly the same, however, and reference can be made to what was written with regard to the latter Secondary infections by C. chrysosperma are often responsible for a further loss of production when it attacks stems and branches already damaged by D. populeum or Septoria musiva, another significant agent of bark necrosis.

The pathogens-The form-genus Cytospora Ehrenb. embraces the anamorphs of many Valsa Fr. species (fam. Valsaceae, ord. Diaporthales, phylum Ascomycota), including facultative parasites very widely found on several broad-leaved trees and characterised by a polyphagy that may be more or less marked according to the species.

C. chrysosperma (Pers.: Fr.) Fr. (teleomorph: V. sordida Nitschke) only infects Salicaceae. It is of little consequence in natural stands, but of primary significance in intensive poplar growing, where its opportunism is favoured by the application of slack cultivation criteria. Found almost everywhere in the world, its highest incidence is in central and southern Italy, eastern Europe, the Near East, northern India (mainly in plantations) and in west-central U.S.A. (especially in Colorado, where it also attacks natural formations of P. tremuloides). Its poplar hosts belong to the Aigeiros, Tacamahaca and Leuce sections (mainly on the just cited P. tremuloides), though its virulence sometimes varies from one isolate to another.

The polyphagous C. ambiens Sacc. (teleomorph: V. ambiens Sacc.) and C. nivea (Hoffm.) Sacc. [teleomorph: Leucostoma niveum (Hoffm.: Fr.) Höhn. (= V. nivea Fr.)] are more sporadic on poplars, though the latter is not uncommonly observed in southern Africa.

As already mentioned, these three species give rise to distinct symptoms on poplars and differ in some morphological and/or morphometric features of their conidia and spores; their conidiomata, however, are always plurilocular (i.e. their interior is divided into some compartments sharing a single orifice), and their conidia are always one-celled, hyaline, cylindrical and curved.

Biology and relations with the host-The establishment of Cytosporae on the host, their fructification times and patterns, the dissemination of their conidia and the progression of the cankers they cause are very similar to those of Discosporium populeum.

Unlike the latter, however, the teleomorph rarely appears on infected tissues and is of virtually no phytopathological consequence. C. chrysosperma has also been shown to start latent infections in buds, that then extend to the bark in case the plant becomes weak.

C. chrysosperma is generally less pathogenic than D. populeum, though it is a more powerful saprophyte that not uncommonly attacks nearly dead adult trees and appears on saplings already invaded by other bark parasites. It is in any event a very plastic organism, capable of vegetating-as shown by its virtual ubiquity-in a very wide range of climates, though it tends to focus on areas subject to drought at least for part of the year, and its effect increases the more a plant is distressed.

Experiments revealed a substantial analogy between the physiological consequences of water shortage and defoliation on the host, consisting in a decrease of carbohydrates in the cortical tissues which, through complex biochemical reactions, causes an impairment of its ability to prevent the establishment of Cytosporae.

Defence mechanisms such as the production of periderm and lignification of the cells near the infection sites are inhibited; direct stimulation of the growth of C. chrysosperma by tissue dehydration, however, seems less likely.

Control strategies-A specific genetic selection of resistant clones is unfeasible against C. ambiens and C. nivea and also seems poorly practicable against C. chrysosperma, since the host's genetic stock is of marginal importance for the growth of the disease by comparison with its physiological state. Nevertheless, it might exert an indirect influence through the environmental adaptability of the various clones, insofar as more constitutionally balanced clones with a harmonious development of the epigeous parts and of the root apparatus, and not excessively directed to productivity alone, will be preferably employed, since they will be less likely to fall into the states of distress that open the way to the disease.

Due to their many affinities, what has been said on the subject of cultivating and chemical control with regard to D. populeum is equally applicable to the Cytosporae. As already mentioned, measures taken to prevent attacks of the former should shield from the latter too.

Canker Caused by Hypoxylon Mammatum

Hypoxylon mammatum is an ascomycete responsible for the most serious disease to which P. tremuloides is exposed. This tree is widely distributed in North America as a component of either pure or mixed forests and is highly regarded as the source of clones suitable for the intensive production of fibre wood in temperate and cold climates. Heavy attacks by H. mammatum have been the cause of heavy economic and ecological repercussions for decades, especially in the U.S.A. on account of the importance of its preferred host, while in recent years there have been growing reports of attacks on other species, especially members of the Leuce section, in several parts of Europe.

Symptoms-Attacks by H. mammatum occur in natural formations, both pure and mixed, of P. tremuloides in the U.S.A. and-to a lesser extent-of P. tremula in Europe, as well as in tall timber stands, especially in those substantially thinned out where the plants keep their lateral branches for a longer period. Infected trees are sometimes more frequent on the edges of a wood, but usually their distribution in the interior is nearly casual, which means that their closeness to healthy trees does not always increase the probability that these will be infected. Both young and adult trees are attacked, the former on their main stem, the latter mainly on their branches, through which the fungus then may reach the trunk. In fact, cankers are often observed around dead branches, pruning cuts and the holes made by xylophagous insects.

A typical set of symptoms appears on P. tremuloides and P. tremula, with an often deadly outcome. A brown to blackish fluid exudes from the infected portions of the bark, which at first are orange-yellow and slightly sunken compared to the surrounding healthy bark. These areas then crack along necrotised rectangular laminae that eventually fall off and expose the inner cortical tissues and the first woody circles, which are blackened and crumbling. The necrosis moves much more quickly along the axis of the stricken organ, whereas its transverse extension is sometimes checked by the formation of a reaction callus on the part of the host. Cases of regression, however, are extremely rare Old cankers are thus narrow and even more than one metre in length; their centre consists of a rough, blackish surface, surrounded by a cracked strip in the process of peeling off, whereas the borders of further expansion are yellowish. Removal of the bark along these ones reveals a white mycelium arranged like a fan of on the cambium.

From 5 to 14 months after the infection, in a period which runs from spring to the height of summer, a stromatic mass appears under the periderm of the cracked peripheral strip. This supports the fructifications of the anamorph (ascribable to Geniculosporium Chest. et Green) in the form of a dusty, bristle-like grey layer composed of several synnematal conidiomata. It is the growing pressure of this layer on the inner surface of the periderm that results in its peeling away from the innermost tissues, so that the conidia can be dispersed. While the fungus attacks new portions of the trunk, the anamorph fructifications that thus find themselves in a central position disappear little by little. The teleomorph, consisting of perithecial ascomata partly immersed in carbonaceous sexual stromata, appears three years after the infection in the blackened, marcescent areas mentioned earlier.

Incidence and the damage caused-Studies of P. tremuloides stands have shown that, despite the notable incidence of the disease on plants in all diameter classes, younger trees display a proportionally higher mortality rate because the cankers form on their main trunk. Instead, in adult trees the branches are primarily affected, and their foliage is thus weakened due to impairment of the conducting tissues, but they are rarely killed by annular necrosis at their base. This last phenomenon is more likely on the upper zone of the trunk and result in withering of the portions of foliage above it; the plant can, however, survive by activating, as a dominant shoot, a branch or a bud below the necrosis. The marked mortality of the younger trees, on the other hand, especially those less than 9 cm in diameter at breast height, is caused by direct infection of the trunk, whose entire circumference may become necrotic in the space of a few years. Both young and adult plants are also likely to be broken by the wind at the canker sites, since their underlying wood is also degraded by the parasite.

H. mammatum is thus responsible for both quantitative and qualitative damage: economic losses of the order of millions of dollars a year in the time of utilisation are caused in the U.S.A. The data concerning its incidence on P. tremula in Europe are fragmentary. Nevertheless, the frequency of infections was valued on high levels in stands of the southern Alps and the southern Jura where, according to Pinon (1986), up to 10% of trees display signs of disease.

The pathogen: taxonomy and geographical diffusion-Hypoxylon mammatum (Wahl.: Fr.) P. Karst., long cited as H. mammatum (Wahl.) J.H. Miller [= H. pruinatum (Klotz.) Cke.], is a member of the fam. Xylariaceae (ord. Xylariales, phylum Ascomycota). A recent suggestion is that it should be renamed Entoleuca mammata (Wahl.: Fr.) J.D. Rogers et Y.-M. Ju, on account of certain morphological characters regarded as excluding it from the genus Hypoxylon Bull. sensu stricto as this is understood today.

In addition to various poplar species, it has been reported on willow, birch, alder, hornbeam, beech, elm, sorb, pear-tree and spruce. Crossed inoculations of isolates have revealed a certain degree of host specificity, though not enough to justify the creation of separate species, since there are no significant morphological differences with regard to ascomata and ascospores. The pathogenic potential of H. mammatum, however, is only realised to the full on poplars, especially on some members of the Leuce section, namely P. tremula, P. adenopoda (Chinese aspen), P. grandidentata and above all P.tremuloides, and also-to a lesser extent-on P. alba and P. × canescens. Among the Tacamahaca, cases of infection have been reported on P. trichocarpa, with a more or less positive outcome depending on the clone concerned, and on P. balsamifera, though rarely. Hybrid clones can also be attacked if one of their parent lines is susceptible.

The areale of the pathogen has been defined in general terms, though some uncertainties remain. It covers the whole of North America, but the incidence of the disease is only heavy in the Great Lakes Region and the Prairies of the North-West, whereas it is secondary in the central section of the Rocky Mountains and even negligible in their northern section and in Alaska, where yet P. tremuloides is abundant and the fungus is regularly reported, even as an asymptomatic endophyte. Its distribution in Europe is irregular; the specific literature contains reports from much of France (with a high incidence in the southern sectors of the Alps and the Jura, characterised by more irregular precipitations), Andorra, Great Britain, the Swiss Jura, some upland and highland areas in Italy (western Alps and Tuscany), Sweden, the Czech and Slovak republics and some parts of Russia. Although all these reports are rather recent, it is believed that H. mammatum has long been present in Europe and may indeed be autochthonous.

Biology and relations with the host-Many aspects concerning the inoculation of H. mammatum in nature, the infection, the role of the environmental variables and the sources of host resistance are still not entirely clear. The fungus can be said to be a recurrent endophyte of several broad-leaved trees. Its pathogenic potential is only triggered by the concomitance of some circumstances, some of which are under specific genetic control. This would explain why the induction of the disease is restricted to some poplar species, and indicates that the real geographic distribution of the fungus is very much wider than that documented by the canker appearance.

The relative importance of conidia, myceliar fragments and ascospores in the diffusion of the disease was very doubtful until a few years ago, but it is now considered that ascospores are the most important. Brown and ellipsoid (20-34 × 9-14 μm), one-celled, they are produced inside globose perithecial ascomata (Ø = 1.5 mm), covered by a greyish efflorescence in the young stage, assembled in groups of 4 to 30 in black and coriaceous stromata. These ascomata contain paraphyses (specialised hyphae that facilitate dissemination) and cylindrical asci (140-200 × 12-16 μm), which pass through the ostiole one at a time and release the ascospores one by one whenever the stroma is wet and the temperature near the bark is above freezing, usually at the end of autumn or the beginning of spring. If the humidity is sufficient, dissemination proceeds according to a 12-hour cycle, taking advantage of strong and weak currents of air for the horizontal and vertical dispersion of ascospores respectively.

The disease development is not dependent on weakness of the host; in any case, the susceptibility of this one to infections is determined by its genetic stock (at least in terms of sensitivity to phytotoxins produced by the parasite and of reaction to colonisation), by environmental and/or cultivation conditions, by the presence of animals responsible for the creation of potential infection sites. It seems certain that penetration, at least of the main trunk, requires a breach, however tiny, in the bark so that the cambium or the xylem is exposed. In fact, the germination of ascospores is prevented by both the green layer and the secondary phloem. The outer woody tissues, however, are colonised first, whereas the cambium and the cortical layers are only killed in a second stage.

This prevention is due to the prior production by the host of fungistatic compounds (cathecol, salicin, salicortin), and also-though there is no consensus on this point-by an ex novo synthesis, stimulated by the presence of the fungus, of low concentrations of other more specific inhibiting compounds (phytoalexins). Even mild water stress, not necessarily demonstrated by the plant with manifest decline, assists infection in two ways, namely by depressing the production of the aforesaid inhibiting substances and by increasing the concentration of some aminoacids (proline, alanine, glutamine) that stimulate myceliar growth. It is also been demonstrated that fertilisers with an abundance of nitrogen compared with phosphorus and potassium encourage the onset of cankers, since plants grow with more succulent tissues, rich in proteins required by the parasite.

In the Great Lakes Region, where the incidence of the disease is highest, a very high percentage of cankers are associated with wounds caused by insects, such as the oviposition scars made by Rhynchota and especially the galls formed after oviposition by wood borers (above all Saperda inornata Say), in whose central tissues ascospores are not exposed to the action of the fungistatic compounds. Penetration may be also assisted by downy woodpeckers (Picoides pubescens L.), which feed on these insects and cause wounds that extend down to the xylem. By contrast, other canker agents frequent in the Rocky Mountains and endophyte fungi in the bark are thought to compete with H. mammatum and thus oppose its infections.

Much uncertainty persists concerning the real pathogenic importance of the phytotoxins (including that with the highest concentration, hymatoxin A), produced in varying amounts by individual fungus isolates. No correlation has been demonstrated between the sensitivity of P. tremuloides clones to these toxins and their susceptibility to the onset of cankers, suggesting that these metabolites are secondary determinants, in other words that they contribute to the virulence but are not needed for the establishment of the disease.

Lastly, some ambiguity is attached to the function of the one-celled, hyaline and fusiform (2-3 × 5.5-6.7 μm) conidia, which are omnipresent in nature where there are cankers, but have a very limited ability to give rise to colonies. It has been suggested, though not yet demonstrated, that they are involved as spermatia in the sexuality of the fungus.

Control strategies-One essential preventive measure is the avoidance of exchanges of wood infected by H. mammatum, even between countries where the fungus is already present, since this could lead to the reciprocal introduction of strains that might prove more virulent in contexts differing from those of their origin with regard to the environment and the genetic stock of their host, as well as the latter's associated mycoflora and its interactions with the same pathogen.

Exclusive control must obviously be accompanied by the search for the most resistant genotypes. This can be greatly accelerated by inoculation tests on saplings, which have proved reliable and in the space of a few days can indicate the rate of lignin deposition in the cell walls and the ability to form callus tissue in response to the infection, both of which are characteristics associated with a plant's susceptibility in the field. Genetic selection, indeed, is the indispensable prelude to the extended use of Leuce species for the intensive production of fibre wood in forest stations or on hydromorphous soils.

Chemical control seems out of the question in situations where attacks by H. mammatum are recurrent, and sanitation measures, designed to wipe out the disease from a stricken station, have proved to be fruitless. P. tremuloides stands with more than 25% of their trees infected must be felled as soon as they reach an appropriate mean diameter, and then replaced with other species or at least with genotypes whose resistance has been established.

The incidence of the disease can be substantially limited through a proper silvicultural management, which must provide for few, slight thinnings, found favourable to infections, and the attainment of fully stocked uniform stands.

Necroses and Cankers Caused by other Pathogens

Other pathogens infect both the trunk and branches, and sometimes result in combined attacks by different entities on the same poplar districts.

These pathogens are not presented here as a subordinate group because they are of secondary phytopathological importance, but because they occur in particular contexts: some are wound or secondary parasites that are present in a latent state in most poplar cultivations and only launch an attack in certain growing and/or environmental conditions; others have a currently limited areale, but are strongly pathogenic and responsible for considerable damage in the involved regions.

Cankers caused by Phomopsis spp.-Necroses or cankers caused by fungi belonging to the form-genus Phomopsis (Sacc.) Bubák were reported in all areas where poplars are intensively cultivated, more frequently in the warm-temperate and sub-tropical zones, including those where poplar growing have been recently introduced, such as southern Africa or India. Common features of these reports, whether remote (in Italy and Germany since the beginning of the century) or contemporary, are the rarity or absence of the teleomorphs-which are ascribed to Diaporthe Nitschke (fam. Valsaceae, ord. Diaporthales, phylum Ascomycota), with black perithecial ascomata and two-celled hyaline ascospores-and an occasional recurrence of these parasites in nurseries in precarious conditions or subject to water stresses. The pathogenic activity of individual species seem fluctuating in certain areas: long periods in which the parasite is apparently absent alternate with others marked by a considerable incidence, sometimes during or after a dry period.

As weakness parasites, the poplar Phomopsides attack more or less debilitated saplings with at least partially dehydrated cortical tissues, via small breaches in the bark itself or leaf scars. The cultivation conditions in which they act-almost solely in the nursery-are similar to those described for the more pathogenic Discosporium populeum, from which they also differ in the limitation of their infections to very young shoots or to the less lignified apical portions of the trunk and the branches. In temperate climates, the hosts are more susceptible during their winter rest, due to their reduced ability to oppose infections and to the lower tissue hydration, often typical of this phase.

The symptoms of attacks by Phomopsis spp. are much the same as those induced by D. populeum; diagnosis must thus be based on examination of some of the macroscopic and microscopic characters of the anamorph fruiting bodies. The pycnidial conidiomata of Phomopsis spp. (with the exception of P. macrospora) are smaller than those of D. populeum and conidia are emitted in mucilaginous cirri that are paler than those of Cytospora chrysosperma, which is also widely diffused in warm-temperate areas and causes similar symptoms. A peculiar feature is the production of two differently shaped one-celled, hyaline conidia: those of shape A are from ovoid to fusiform, those of shape B are filamentous, straight or curved. The individual species, especially those encountered in Europe, are mainly distinguished in the light of their morphometric characters; uncertainties persist, however, and the employment of modern biomolecular diagnostic techniques would be advisable. The following species were found on poplars:

- P. putator (Sacc.) v. Höhn., reported at the start of the twentieth century in Germany and more recently in Italy and Argentina, on Aigeiros hybrids;
- P. populina Vogl., reported in the past in Italy on Aigeiros clones of the so-called "Canadian" type;
- P. pallida (Fck.) Sacc. et D. Sacc., reported in Portugal and Italy from the 1940s to the 196Os as mainly virulent on Populus deltoides, but also on Populus alba;
- P. tirrenica Moriondo, reported on Euramerican clones (including "I-214") in Italy since the early 1960s;
- P. macrospora Kobayashi et Chiba, reported in Japan and present with more continuity in the U.S.A. along a belt running from Minnesota to

Mississippi, both in nurseries and plantations up to three years of age, on Populus deltoides, Populus × euramericana (including clone "Robusta"), Populus nigra and Populus maximowiczii.

An effective control of these parasites, too, must be primarily based on the application of careful cultural practices, designed to prevent saplings from damage and dehydration. Pruning residues are readily colonised by Phomopsides and their removal is an important step towards the extinction of sources of inoculum.

Necroses and cankers caused by Fusarium spp.-Various species of the form-genus Fusarium Link-whose teleomorphs are referable to Gibberella Sacc. (fam. Hypocreaceae, ord. Hypocreales, phylum Ascomycota), rarely seen on infected hosts-cause characteristic cankers, sometimes called fusarioses, on poplars from various sections. They are all markedly polyphagous, both on poplars and on other broad-leaved trees and herbaceous plants in the temperate zones. Geographical distinctions with regard to Populus thus are aleatory. avenaceum (Fr.) Sacc. is perhaps the most important species at present. First reported in the 1950s on Euramerican clones in France, it has since spread to very different parts of Europe, ranging from central and eastern areas with a continental climate to sub-mediterranean areas and recently to Portugal with its oceanic climate. F. solani (Martius) Sacc. emend. Snyder et Hansen, found on Aigeiros and Tacamahaca poplars and intersectional hybrids, originally seemed to be confined to North America, but then has been reported in Poland too. Sporadic and of limited importance are F. lateritium Nees, observed in some parts of France and in the U.S.A. on P. trichocarpa, and F. sporotrichioides Sherbakoff, reported in eastern Europe and in central Italy on P. × euramericana.

The genetic component is of little consequence in fusarioses. The predisposing factor is distress of the host due to a shortage of water or poor growing conditions, such as the excessive crowding typical of many nurseries. Nevertheless, the inoculation seems to require discontinuities or points of greater vulnerability in the bark, preferably at the boundary between two vegetation stages or at shear planes. Some workers have suggested that latent infection of tissues surrounding the buds can also take place. Laboratory experiments have shown that the fungus is thermophilous to a certain extent, and this explains why its host is most susceptible in spring, when vegetative resumption is at its height, by contrast with the attacks of other bark parasites.

The incidence of the Fusaria is not comparable with that of Discosporium populeum in absolute terms. They are, in fact, almost solely a cause of concern in nurseries, especially on first-year and sometimes second-year saplings, where their initial appearance on a few plants may be quickly followed by involvement of the entire stand. Attacks on 10-to 15-year-old adult trees by F. avenaceum has occasionally been reported; in any case not more than 10% of the plantation is affected, while the symptoms are not immediately visible and mostly take the form of disorganisation of the cortical tissues in a preferential belt of the trunk 4 m to 8 m in height from the ground.

In the nursery, F. avenaceum is responsible for a characteristic set of symptoms. Swellings of the infected areas, often at the transition ring between one vegetation stage and the next, are soon followed by numerous close together longitudinal fissures, known to growers as "cat scratches", that advance in step with the necrosis of tissues and to the point where the bark almost begins to fray. During the late spring and summer, these lacerated areas take on a typical reddish to purple colour, typical of F. avenaceum, due to the presence of its mycelium. Cushion-shaped necroses with a different appearance sometimes form around the buds, on which fruiting bodies of the anamorph (sporodochial conidiomata) differentiate in conditions of high atmospheric humidity. The elongated, sickle-shaped conidia (14-40 × 3-5 μm), usually divided by 3-4 septa, are released in mucilaginous masses that are subsequently diluted by the rain. During the vegetative season, the stricken saplings almost always succeed in halting the further spreading of the necroses through the formation of evident callus tissues, showing thus the cankerous stage of the disease. Although the survival is assured, their marketability is greatly impaired, since these callus zones, within which the parasite has in any event disorganised the initial woody cell layers, are preferential sites for breakage even on the part of moderate winds, as well as being vulnerable to attacks by other bark parasites also active during a plantation's early years, such as D. populeum and Cytospora chrysosperma.

The symptoms of an attack by F. sporotrichioides are very similar, whereas F. solani causes the appearance of brownish necroses that are accentuated by the concomitant action of other Fusaria and of Geotrichum Link species. In north-eastern Europe it has been observed that pathogenic strains of F. solani can also be responsible for the degeneration of hyphae and conidia of Ceratocystis fimbriata, another poplar pathogen that causes what is known as "black canker" or "target canker", and that both fungi are sometimes isolated from lesions of this type.

As can be said with regard to the other bark parasites, the control of Fusaria first of all is founded on good nursery practices. Cuttings must be protected from physiological stresses as far as possible, especially with respect to the high reciprocal competition imposed by an excessive density, and care must be taken to avoid damaging their bark. The genetic selection of possible resistant clones is not for the present a viable solution, since the degree of susceptibility of existing clones is not yet well known and the relative weight of the genetic component compared with the environment is still uncertain. Chemical control is about the same as for D. populeum.

Sooty-bark canker caused by Encoelia pruinosa-Encoelia pruinosa (Ell. et Ev.) Torkelsen et Eckblad [= Phibalis pruinosa (Ell. et Ev.) Kohn et Korf = Cenangium singulare (Rehm) Davidson et Cash] is a discomycete (fam. Leotiaceae, ord. Leotiales, phylum Ascomycota) responsible for a serious canker on P. tremuloides and, to a very secondary extent, on P. balsamifera. The disease involves a very wide geographic belt, similar to the areale of P. tremuloides natural forests, running from Alaska through western Canada and along the Rocky Mountains, with extensions to some of the Mid-west States, down to northern Mexico. Besides, the pathogen was observed in Norway. pruinosa is a wound parasite and thus easily able to attack P. tremuloides, whose soft bark is readily damaged by climatic factors, animals and humans. Recent surveys in natural formations have shown that its incidence on all live plants is low (of the order of 1%), but much higher when only mature and overmature trees (aged 100 to 120 years) are considered, contrary to what is observed with Cryptosphaeria lignyota (Fr.: Fr.) Auersw., another bark canker agent that attacks this host

The symptoms are rather characteristic: colonisation is revealed by an elliptical area depressed with respect to the still healthy surrounding bark, arranged around the infection site, in rapid extension throughout the year without the host being able to form an effective callus of reaction. The dead tissues become more striking after 2-3 years, when the pale outer layer of the bark begins to peel off and expose the thicker inner layer, which is blackish and remains tightly attached to the trunk even for a long time after the plant is dead. When touched, its surface crumbles and leaves a sooty deposit on the fingers, hence the common name of the disease. The fungus seems to cement this layer to the wood below, to which it confers a design composed of dark spots on a paler background, visible when the same layer finally falls off in long fibrous strips.

By contrast with most bark parasites, only the teleomorph of E. pruinosa has so far been found in the field. This consists of externally pale grey apothecial ascomata in the shape of irregular cups (Ø = 3 mm), produced in large numbers on the old blackened bark or under the thin whitish outer layer. The ascospores are primarily disseminated by the wind all through the year, whenever the temperature and humidity are right. In the laboratory a conidial form has been obtained, variable according to the medium and growth conditions, and comparable at times with Acremonium Link and at times with Myrioconium Syd., whose importance in nature, if any, remains to be determined.

Sooty canker is a serious disease on P. tremuloides, especially on large trees, mainly because its appearance does not leave any way of escape for those that are stricken and their chances of mounting a successful reaction are very small. Statistical analysis has shown that the mean expansion rate of a typical canker lesion around the circumference of the trunk is about 16 cm/yr, which is enough to kill a tree in a few years. It is estimated that more than half of the deaths in natural and nature-like forests of this poplar species are directly caused by E. pruinosa.

Snake canker and wood decay caused by Cryptosphaeria lignyota-Like Encoelia pruinosa, Cryptosphaeria lignyota (Fr.: Fr.) Auersw. [= C. populina (Pers.) Sacc.] (fam. Diatrypaceae, ord. Diatrypales, phylum Ascomycota) is a pathogen especially present on P. tremuloides, within an area of activity that runs from Alaska and Yukon Territories along the western U.S.A. down to Arizona and New Mexico. It has also been reported in some central and eastern States, e.g. Minnesota, Michigan and Pennsylvania, where it may attack other poplar species (P. balsamifera, P. trichocarpa, P. deltoides), as well as willow (Salix purpurea L.). At present, its incidence is confined to North America; further epidemiological developments, however, may be foreshadowed considering that more than a century ago the fungus was found in Europe on P. tremula, P. alba and P. nigra.

lignyota is a canker agent that mainly infects young P. tremuloides trees in natural formations, with a diameter at breast height of not more than 30 cm, and more frequently the younger they are. It can also infect buds and branches and then pass to the main trunk to induce a typical canker syndrome. More mature trees are not affected by the disease, possibly because the fungus is in some way limited in both its penetration of very thick barks and the expression of symptoms on the same. The pathogen is important not only as a bark parasite, but also on account of its ability to reach the central cylinder of even mature trees and behave like a wood decay agent.

The penetration stage requires a breach in the bark or, otherwise, the infection may also spread from previously stricken organs of the plant. The cankers are usually long and narrow, sometimes coiled around and along the trunk like a snake due to the marked difference between their longitudinal and their lateral progression, surrounded by areas of ochreish or orange bark, symptomatic of the advancing colonisation. Callus formation does not become apparent for at least two years after infection and is often unable to prevent its further extension. Bark tissues dead for at least a year, tightly attached to the sapwood, become black and fibrous, like those found in E. pruinosa cankers, from which they are distinguished by the presence of scattered elliptical pale spots (Ø = 0.5-2 mm). The perithecial ascomata typical of the teleomorph develop in a very compact, narrow myceliar matrix (pseudostroma) that works its way under the periderm of this dead bark and may be up to 30 cm long. In some instances, acervular conidiomata of the anamorph differentiate on the edge of the cankers; these are pale orange and produce one-celled, filiform and curved conidia assigned to Libertella Desm.

The anamorph is often isolated as a decay agent from both the heartwood and the sapwood, where it induces various colourings: grey in the sapwood under the cankers of thinner trunks, from brown to yellow and on to pink with mottles in more mature trees.

The parasite can kill young trees within a year from infection and contributes to the natural thinning of P. tremuloides formations. It would seem that often the host's death is not due to annular extension of the canker, but to a massive invasion of the sapwood that prevents it from transporting water and nutrients. Owing to its marked preference for shoots and young trees, the wood decay it causes and its proven ability to establish itself on a variety of poplar species, C. lignyota is also a potential threat to intensive poplar growing, and strict vigilance is essential to avoid its introduction.

Black or target canker caused by Ceratocystis fimbriata-Several species of Ceratocystis Ell. et Halst. (ord. Microascales, phylum Ascomycota) were reported in association with cankers on poplars.

The only one of remarkable phytopathological importance is C. fimbriata Ell. et Halst. [= Endoconidiophora fimbriata (Ell. et Halst.) Davids.], a parasite very widely distributed in temperate and tropical environments and on the whole very polyphagous, since it is able to attack roots, stems or fruits of both herbaceous plants (sweet potato, cacao, coffee) and trees (various species of Prunus L., rubber trees and above all planes, on which, with the f. sp. platani Walter, it causes a pernicious tracheomycosis known as "black stain"). On poplars, too, its range of action is being gradually extended. Up to twenty years ago, its manifestations were confined to Alaska and the Rocky Mountains, where it is one of the main agents of cankers in natural P. tremuloides stands-rarely fatal on the other hand-, and more sporadically to a few of the north-eastern U.S.A. and Quebec. Attacks of a certain seriousness have since been reported on adult plantations in Poland, especially on some Tacamahaca, Tacamahaca × Aigeiros and Aigeiros × Tacamahaca hybrid clones, and more recently, though with a secondary incidence, on P. deltoides in India. fimbriata penetrates into the trunk through the insertion points of small branches, as well as through wounds, frequently opened in the tender thin bark of P. tremuloides. In the forest, it mainly attacks trees with a diameter of about 40 cm, but not the older ones. The symptoms are similar on P. tremuloides and on the hybrids mentioned above: the first sign is a slight depression in the bark, which gradually turns brown starting from the infection point and, in the following spring, is circumscribed by a vigorous callus. During the subsequent vegetative repose, this too is colonised and killed by the parasite, resulting in the formation of other callus tissue around the first when vegetation resumes. Repetition of this annual process leads to the creation of a large target-shaped canker, with sinuous borders but an approximately ellipsoid shape, formed of a cracked central concavity surrounded by blackish, concentric callus folds, which is similar to cankers caused by Nectria galligena Bres. on P. tremuloides. After some years, the degraded cortical tissues peel to reveal rings of dead xylem, but the death of a tree is a rare event.

Secondary symptoms, occasionally observed on the foliage of P. tremuloides, take the form of blackish angular notches that preferably form near the vascular bundles, and of a necrosis of young shoots following the passage of the fungus via the petioles.

The typical teleomorph fruiting bodies form on the edges of cankers, in tissue dead since at least one year. They are black perithecial ascomata that burst out on maturity, with a characteristic long neck, and whose surface sometimes bristles with conidiophores (Table...). The one-celled, hyaline and hat-shaped ascospores are released in mucilaginous masses and can be disseminated by meteoric agents or insects (usually sap beetles of genus Epuraea Er.). The anamorph is ascribed to Chalara (Corda) Rabenh. (previously assigned to Thielaviopsis Went) and has three one-celled conidial forms, generated by single conidiophores or grouped in synnematal conidiomata: hyaline cylindrical conidia (8.5-32.8 × 2.7-7.5 μm), hyaline or brown barrel-shaped conidia (6.4-13.0 × 4.0-9.6 μm), and brown, thick-walled ovoid conidia (7.0-24.0 × 5.1-15.2 μm). They are produced on the infected bark or (especially the cylindrical type) in the underlying vascular tissues.

The parasite colonises the phloem and the xylem below the infected bark, both inter-and intracellularly, in which case it spreads through both the pits and the perforated plates. Resistant genotypes block further diffusion of the fungus at an early stage by differentiating cork and callus layers at the edges of the infected zone, together with cells containing tannic substances with an inhibiting action. The resistance response would also seem to be correlated with the concentration of growth hormones (auxins, cytokinins) in the plant, though the nature of this link has not yet been established. Inoculation tests with isolates from other broad-leaved plants on susceptible poplar clones, resulting in a limited degree of development of their cankers, revealed a certain, though not absolute, specificity in the relationship with the host.

Other Ceratocystis species, all encountered sporadically and almost solely on P. tremuloides stands in the Rocky Mountains and Alaska and of negligible phytopathological importance, are associated with cankers or dead bark zones. The possibility that some may be secondary invaders of cankers already produced by C. fimbriata cannot be ruled out.

Other agents of cankers or bark alterations on poplars, whose incidence is more local and/or secondary, are the following:

- Nectria galligena Bres. (fam. Hypocreaceae, ord. Hypocreales, phylum Ascomycota), with anamorph Cylindrocarpon mali (Allesch.) Wollenw., a polyphagous parasite widely distributed throughout the northern temperate zone on fruit trees and forest trees, observed sporadically on poplar mainly in central Europe (Germany, Hungary) and in the Rocky Mountains; in this region, it can cause on P. tremuloides cankers with very evident calluses that can only be distinguished from those of Ceratocystis fimbriata by isolation of the fungus or by the appearance of its reddish fruiting bodies on cankers;
- Rhytidiella moriformis Zalasky and R. baranyayi Funk et Zalasky (fam. Cucurbitariaceae, ord. Dothideales, phylum Ascomycota), agents of a syndrome known as rough bark or cork bark on P. balsamifera and P. tremuloides respectively in western and central Canada; both are responsible for a gradual thickening of the entire bark of trunks and branches more than two years old, due to the death of the surface layers and abnormal proliferation of the underlying phloem, with the formation of corky ridges separated by deep fissures and progression to a greyish colour; R. moriformis also has an anamorph referable to Phaeoseptoria Speg., which appears on two-to five-year-old bark, followed by the teleomorph two or three years later;
- Diplodia tumefaciens (Shear) Zalasky, the anamorph of Keissleriella emergens (Karst.) Bose (fam. Lophiostomataceae, ord. Dothideales, phylum Ascomycota), which wasreported on P. tremula in northern Europe but is much more widespread in Canada and the northern U.S.A. on P. tremuloides and P. balsamifera; by contrast with the Rhytidiellae, it attacks small areas of the bark, where it causes an alteration similar to rough bark, and may also infect one-year-old branchlets and the main roots near the trunk (where the teleomorph

sometimes differentiates); an additional symptom is the formation of distinct woody galls, due to hypertrophies of the bark and of the sapwood induced by the fungus;

- Phoma exigua Desm. var. populi de Gruyter et Scheer (Mitosporic Fungi), an opportunistic parasite observed in the Netherlands in the 1990s and also the subject of earlier reports from Germany and Italy referred at the time to Phoma urens Ell. et Ev.; favoured by mild, damp winters, it causes cortical lesions on shoots and one-year-old branchlets of some P. nigra, P. × euramericana, P. × interamericana and P. trichocarpa clones which are quite similar to those provoked by Discosporium populeum, it is distinguishable, however, by the aggregated or confluent pycnidial condiomata and by their conidia, that are in part two-celled;
- Corticium salmonicolor Berk. et Br. (fam. Corticiaceae, ord. Stereales, class Basidiomycetes), a polyphagous species found in the tropics as the agent of so-called "pink disease" on eucalyptus, rubber tree, tea shrubs and other cultivated plants; in India, its incidence has recently been increasing in P. deltoides, P. × euramericana and P. yunnanensis Dode plantations exposed to high temperatures and humidity; its attacks on the trunk and branches often kill the main shoot;
- Dothiorella gregaria Sacc., the anamorph of Botryosphaeria dothidea (= B. ribis Gross et Dugg; fam. Botryosphaeriaceae, ord. Dothideales, phylum Ascomycota), an extremely polyphagous species (found on more than 100 genera of shrubs and both broad-leaved and coniferous trees) in the temperate zones and the tropics; it is able to attack plants already weakened by water stress or other diseases; on poplars, it is mainly active in north-central China, where it causes a bark necrosis on the trunk and branches, known as blister canker or ulcer disease, that develops rapidly in the spring and autumn and results in heavy production losses in the plantation;
- Botryodiplodia populea Z.K. Zong (Mitosporic Fungi), that has been observed over the last ten years in China, where it causes cankers on various poplar species, though its incidence is less than that of D. gregaria.

DISEASES CAUSED BY SEPTORIA SPP.

Septoria musiva Peck is a leading bark canker agent and one of the pathogens for which the European Union has promulgated quarantine measures designed to prevent its introduction into the Old Continent. Nevertheless, it can also attack the foliage; besides, the approximately ten other Septoria Sacc. species reported on poplars, all with teleomorphs referable to Mycosphaerella Johan. (fam. Mycosphaerellaceae, ord. Dothideales, phylum Ascomycota), only attack leaves.

In view of the composite symptomatology induced by the species of greatest interest, the Septoriae are here treated separately, including a few mentions on the most important leaf parasites.

Cankers and Leaf Spots Caused by Septoria Musiva

Symptoms and the damage caused-S. musiva attacks young poplars in nurseries, coppices for the production of biomass, windbreaks in the North-American Plains, as well as in intensive plantations for the production of wood. Although its compatibility with the host is in part genetically controlled, distress of the plant seems to be a predisposing factor for infection, though not as decisively as for Discosporium popoleum.

Attacks begin in late spring and continue throughout the summer. The initial signs are irregular brown leaf lesions, with a whitish or yellowish centre, that are usually larger than those caused by Septoria populicola Peck and eventually join together to kill extensive portions of tissue. These symptoms at first appear on the lower foliage, but soon spread to the whole crown and cause moderate defoliation.

The disease goes no further on many poplar genotypes which are autochthonous of North America, and does not extend to other organs. On various introduced hybrids (not of P. × euramericana), instead, it involves the cortical tissues of current-year shoots and new branches, after penetration of the parasite through discontinuities in the tegument (leaf scars, lenticels, wounds) or connections with the leaves (petioles). The infected areas, which are visible owing to their brown colour and their sunken surface compared to the surrounding healthy bark, crack as the result of dehydration, while swellings and calluses form at their edges as the host seeks to stop the necrosis from spreading. If this reaction fails to prevent annular colonisation of the bark, the parts of the plant above succumb; even if it is successful, the large cankers thus created are sources of new inoculum and are often invaded by weakness parasites, as well as being sites of poor mechanical resistance.

In the plantations where attacks of S. musiva are recurrent, the many bark cankers induced by the same and by secondary parasites, together with inevitable wind breakage, generally result in the death of the trees within four years; according to some estimates, the loss of biomass on the part of very susceptible clones may exceed 60%. Stricken nursery plants are rendered unmarketable.

The pathogen-Septoria musiva Peck is the anamorph of the ascomycete Mycosphaerella populorum Thompson.

It spread from its original areale, including north-central U.S.A. (North and South Dakota, Minnesota, Wisconsin, Michigan)-where the resistant genotypes are more frequent but intense attacks are still constant, especially against those clones that have since been introduced-, to the whole of the central and east of North America, as shown by reports from Mexico and some Canadian States. Observed more than twenty years ago in the Crimea and the Caucasus, it has acquired a considerable incidence on wide poplar regions in Argentina, but has not yet appeared in Europe, Africa and Oceania.

Species belonging to the Aigeiros, Tacamahaca and Leuce (Trepidae only) sections are attacked by S. musiva in various ways:

- with few exceptions, P. nigra and P. × euramericana are resistant to the disease, both in the form of leaf spots and bark cankers;
- P. tremuloides is resistant, but its leaves are occasionally attacked (as reported in Mexico);
- P. deltoides, P. balsamifera and P. tacamahaca are moderately susceptible to the leaf lesions;
- several Aigeiros × Tacamahaca hybrids are susceptible to bark cankers, which become limiting on P. nigra × P. trichocarpa, P. nigra × P. laurifolia and especially P. deltoides × P. trichocarpa clones.

Recent studies have found evidence of different levels of aggressiveness in S. musiva populations; a statistical support for the identification of true pathotypes with specific degrees of virulence, however, has not yet been produced.

Biology and relations with the host-In the form of cortical parasite, S. musiva requires breaches in the bark for its penetration, on condition that the potential host is young. Distress on the part of the host due to stresses of various kinds, especially drought and ozone pollution, while not decisive, is a predisposing factor to the appearance and the evolution of cankers. On the contrary, ageing seems to confer resistance, as it is supported by the disease incidence in the field, which is almost exclusive in nurseries and recent plantations.

Even on susceptible young trees, new cankers form only on each year's shoots and not on the two-year-old branches, which indeed appear colonised, sometimes heavily, by Smusiva, but after infections started the previous year. On these older cankers, attacks of secondary parasites, e.g. Cytospora spp. and Fusarium spp., are often observed.

Sporulation is thus entrusted to the mycelium in colonised leaves and freshly formed cankers, where small blackish pycnidial conidiomata differentiate, each containing hyaline and elongated conidia (17-56 × 3-4 μm), with 6 or 7 septa, that are then released in whitish mucilaginous cirri and dispersed by the rain.

During the winter, the teleomorph appears on fallen leaves and sometimes on overwintering cankers in the form of blackish pseudothecial ascomata, in which cylindrical asci (51-73 × 12-17 μm) are produced, each containing eight hyaline uniseptate or pluriseptate ascospores (15-27 × 4-6 μm). Their release late in the following spring causes the primary infections, which seems to be confined to the leaves; however, it is possible that a certain share of these infections is due to conidia overwintered in the cankers. One or two weeks after the appearance of the first leaf lesions, the new conidia ripen and bring about the secondary infections on other leaves and on the sprouting shoots.

Control strategies-Countries in which the accidental introduction of S. musiva is a potential threat must apply severe quarantine measures and set strict limits on the exchanging of biological material. There is a real danger in the case of north-central Europe, where much use is made of many of the Aigeiros × Tacamahaca hybrids that have proved susceptible in North America, whereas in Spain, Italy and the Balkans the prevalent species is the more resistant P. × euramericana.

With reference to this, as for the other pathogens, selection of resistant clones must take due account of their possible susceptibility to other diseases now widespread: particularly, many Euramerican genotypes are exposed to attacks by the leaf parasite Marssonina brunnea, whereas the intersectional hybrids just mentioned would be better chosen in the effective control of this pathogen.

All the agronomic measures indicated against Discosporium populeum are equally effective for S. musiva, even though their adoption does not totally ensure freedom from the leaf syndrome and the eventual transition to bark canker, when the fungus passes to some branches via the petioles. In nurseries where heavy attacks are likely, it is thus advisable to apply a chemical control. Three spring-summer treatments with benomyl-based products considerably reduce the incidence of the disease.

Biological control with the mitosporic fungus Phaeotheca dimorphospora Des Rochers et Ouellette seems a promising prospect, since preventive spraying of both conidial suspensions and culture filtrates has proved an effective way of limiting the onset and progress of cankers.

Leaf Spots Caused by Septoria Spp.

Two of the species that only attack leaves are of particular phytopathological importance:

_ S. populicola Peck (teleomorph: Mycosphaerella populicola Thompson), a leaf spot agent with a considerable incidence in North America, where its attacks sometimes overlaps those of S. musiva; it has recently been reported in southern Africa, though not as responsible for substantial damage;

_ S. populi Desm. [teleomorph: M. populi (Auersw.) Schroet.], the equivalent of S. populicola in Europe (mainly reported in France and Russia), but also observed sporadically in Asia (Turkey, Iran, India), U.S.A. and Argentina.

These two species differ from each other and from S. musiva in some microscopic characters, in their hosts and in some peculiarities of the symptoms they cause. Both seem less polyphagous than S. musiva: S. populicola attacks P. trichocarpa and P. tacamahaca only, and S. populi is observed on P. nigra and P. × euramericana, though in India it was also reported on P. ciliata and P. alba.

The Septoriae require rather high temperatures and humidity, therefore they usually appear in late spring or early summer in nurseries, mini-rotations for the production of biomass or young plantations. S. populicola and S. populi induce a syndrome quite similar to the leaf symptoms described for S. musiva, except that the spots they cause are generally a little smaller; those provoked by the European species are also slightly paler and concave with respect to the surrounding leaf blade. Severe attacks result in yellowing of the crown and premature defoliation.

ROOT ROT CAUSED BY ROSELLINIA NECATRIX

Rosellinia necatrix is responsible for what is sometimes called white root rot, or "pourridié laineux" in Francophone literature. Its infections on poplars has so far been confined to rather limited areas. Nevertheless, since the parasite is common on other arboreous and herbaceous plants-which are possible source of inoculum-and some current trends, mainly concerning cultural practices, could be favourable to it, special measures should thus be taken to prevent its appearance in till now free regions.

Symptoms and the damage caused-In poplar growing, substantial attacks by R. necatrix was only reported in intensive plantations under warm-temperate or sub-tropical climates, especially in those placed on soils with a fluctuating water table or which are anyhow conditioned by an irregular supply of water. Young trees until the second year after planting out and mature or senescent trees are most vulnerable, but infection of seedlings and saplings sometimes occurs too.

When saplings are planted out, the parasite is often already present in the soil as a saprophyte, usually as a carry-over on wood residues left when a previous plantation was felled. Infection by such a dense inoculum is highly probable, since the host is in a state of crisis owing to the transplanting and its roots are small, rich in reserve substances and devoid of protective tissues. The most serious effects appear within 3 or 4 years and often result in the apoplectic death of the entire tree.

During a subsequent phase of the plantation, colonisation and degradation of root cortical tissues by the fungus is usually asymptomatic. This stage may last for several years in chronologically intermediate stands in full vegetative vigour, since their trees manage to circumscribe and/or tolerate the infections by forming reaction tissues and putting out new healthy roots near the uninfected areas of the collar.

When a plantation comes to maturity, which is negatively conditioned by the reduced spacing of trees through an increasing competition for nutrients on the part of their extended and interwoven roots, their ability to regenerate tissues diminishes.

Therefore, invasion of ever larger parts of the root system makes itself evident in the form of stunted vegetation and increasingly extensive yellowing of large sectors of the crown; leaves eventually wither and fall off. The terminal stage of the disease is marked by the death of the main branches dependent on the infected roots and, not rarely, the entire tree is often killed. This gradual decline as opposed to sudden death, however, is also observed in the case of young trees, especially if they belong to clones with good rhizogenic capabilities enabling them to get through the vulnerable stage after planting out and then conquer the disease, at least in appearance.

These epigeal symptoms are non-specific in themselves; they are similar, for example, to those caused by Armillaria mellea (Vahl: Fr.) Kummer, though this usually attacks senescent plants and in different growing conditions. Infections by R. necatrix almost never proceeds beyond the collar, therefore they can be reliably diagnosed by exposing the hypogeal part and looking for a loose, filamentous and embracing mycelium with sections organised in cords or interrupted by plates; it is whitish at first and turns to brown as the rot proceeds. It is composed of bundles of interwoven cylindrical hyphae with diagnostically significant piriform swellings near septa.

The damage is evidently quantitative, since wood growth is stunted as a result of decline and entire trees are lost in young plantations. In some Indians regions, nursery mortality rates of 10-30% were recently reported, whereas fairly recent estimates in Italy have shown on a local scale mortality of as much as 10%, with a loss of almost 1% of the national production.

The pathogen-R. necatrix Prill., long cited as R. necatrix (Hart.) Berl., is a member of the fam. Xylariaceae (ord. Xylariales, phylum Ascomycota) which is able to parasitize hypogeal organs of many herbaceous plants, shrubs and trees, including vine, beet, avocado, various ornamental flowers, broom, elder, walnut, olive, plane, fruit trees, oak, eucalyptus, willow, poplar etc.

Its pathogenic activity on poplars is most marked in Italy, where it has been present for a long time, with an high incidence in areas bounding Po river and near its delta. Here, it was one the main factors that limited poplar growing in the first decades of this century and then, after a period of stasis, in the second half of the 1980s. Considerable attacks have been described during the latest years in Portugal and southern Africa, as well as in India, where are mostly observed in the nursery.

Although all poplar groups are susceptible, certain P. deltoides and P. × euramericana clones seem less affected by infections, not so much because they mount a true resistance reaction as because their tolerance is respectively enhanced by a lower predisposition to water stresses or by a better rhizogenic capacity.

In Italy, other species were occasionally reported on poplar roots in the past, i.e. R. aquila (Fr.: Fr.) De Not. (= Sphaeria aquila Fr.) and R. desmazieresii (Berk. et Br.) Sacc. (= R. quercina Hart.), both of negligible importance on Salicaceae. More recent findings are R. corticium (Schw.: Fr.) Sacc. in France (on branches) and R. subsimilis Karst. et Starb. in Switzerland (on P. tremula branches).

Biology and predisposing factors-The teleomorph of R. necatrix is very rarely found at the base of dead or totally impaired plants. It is characterised by globose stromata (Ø = 1.2-2.0 mm) rising from a short stipe, at first reddish brown to dark brown and then black at maturity, single or in clumps, which erupt from the host cortical tissues. Stromata are initially covered by a felty or woolly mycelial matrix (subiculum) of a similar colour, but then gradually emerge until they are mostly exposed to the air, whereas the subiculum remains as a basal layer of mycelium. Each stroma has an ostiole and contains a single perithecial ascoma (Ø = 1.0-1.5 mm) that opens in ripening to release brownish, one-celled ellipsoid or fusiform ascospores (35.5-40.5 × 6.1-6.9 μm), which are then dispersed by the wind and by insects. The anamorph,

Dematophora necatrix Hart. (= Graphium desmazieresii Sacc.), is much more frequent since it appears, besides on subicula of the teleomorph, on the surface of infected roots. It is composed of flocky synnematal conidiomata, formed by united brown to reddish hyphae (0.5-1.5 mm long) arising from a common base and with dusty hyaline tips due to the formation of conidiophores. These are divergent, light brown, frequently di-and trichotomously branched, and produce light brown globose or ellipsoid conidia (3-5 × 2.5-3 μm).

Propagation of the fungus in poplar plantations, however, seems less dependent on ascospores and conidia than on mycelial fragments installed on woody or herbaceous plant residues, and than on the mycelial cords (similar to true rhizomorphs) that run through the soil from the inoculum sources to susceptible hosts. This usually occurs in spring and autumn, when adequate precipitations and ideal temperatures allow the fungus to spread vigorously as a saprophyte, while its dispersion may be further assisted by local flooding, since the mycelium attached to infected debris can survive for several weeks in running water.

Penetration into the host requires breaches in the bark, which are however numerous in the lower part of the trunk and on the delicate root system of recently outplanted saplings, rich in parenchyma and devoid of protective tissues. The main root near the collar and the more superficial lateral ones are usually the first to be attacked, then the deeper sectors too are gradually involved. When the infection is established, further colonisation is favoured by debilitation of the host, which in warm-temperate areas has often already been caused by a lack of water and sometimes aggravated by epidemics of Marssonina brunnea, a leaf pathogen very active in the summer.

This ability of R. necatrix to propagate as a saphrophyte in wet periods and to invade trees as a parasite in dry periods makes it very insidious when poplars are grown on marginal lands, with coarse soil and a fluctuating water table and often situated near a watercourse, while its remarkable ability of passive survival in adverse conditions is sometimes aided by the organisation of its mycelium into roundish (Ø = 3-4 mm), brown-blackish sclerotia.

Control strategies-As in the case of all broadly polyphagous facultative parasites, ad hoc genetic selection of poplar clones resistant to R. necatrix is impracticable. Nevertheless, it is possible to look for genotypes endowed with a rhizogenic capacity enabling them to tolerate the presence of root rot during the critical post-transplant stage, when the death of plants is more likely.

Priority, however, must be assigned to the adoption of cultivation practices that minimise the probability of attacks, which would not only damage the current plantation, but also those in the future in a dangerous cumulative fashion, since the parasite can survive on the wood debris left on the ground. Since it is not economically feasible to remove all the roots when a plantation is felled, it is right at least to grub up the stumps and take them away without shredding them on the spot, as is often done today in many cultivation districts. As many as possible of the wood residues should also be brought to the surface so that the dry air can impair the survival of the parasite, and the soil should be treated with ammoniacal nitrogen to speed up their biodegradation. Before beginning a new production cycle, the soil needs to be left free from poplars through at least one or two years; in the meanwhile, it is advisable to grow herbaceous plants, e.g. maize, which help in decomposing the biological substrate of R. necatrix if they are appropriately fertilised with nitrogen. This cultivation system considerably reduces the aggressiveness of the parasite, though it extends the woody production cycle. The new plantation should be composed of clones with a good epigeal/hypogeal growth ratio, which react to R. necatrix more strongly. During the cultivation, stressful situations, such as overcrowding, water shortages and leaf diseases, must also be avoided as far as possible, both in the field and obviously in the establishment of a new nursery.

Attempts to treat an already clear attack almost always fail and are only truly beneficial in young plantations, in the case that the distribution of rots is still limited. The most infected trees can be uprooted and the roots of the others can be partly bared to expose the fungus to an unfavourable environment. The initial infection sources can be extinguished by spraying the ground with benomyl suspensions (50 g/plant a.i.) or triforine suspensions (30 g/plant a.i.), in either one or two treatments.

Biological control has nothing concrete to offer at the moment.

OTHER ROOT DISEASES

R. necatrix is the only root rot agent capable of attacking young intensive poplar plantations and impairing the growth and often the survival of plants for much of the cultivation period. Other root parasites are mostly sporadic, though sometimes are endemic in limited regions. They appear in nurseries, plantations, quick-rotation coppices and natural formations; in the last three cases, however, they only infect mature or senescent trees ready for felling and rarely cause substantial damage.

Root and butt rots caused by Armillaria spp.-The genus Armillaria (Fr.: Fr.) Staude (fam. Tricholomataceae, ord. Agaricales, class Basidiomycetes) includes several species variously pathogenic for herbaceous plants and both conifers and broad-leaved trees.

The most widespread and most pathogenic agent is *A. mellea* (Vahl: Fr.) Kummer, long cited as *A. mellea* (Vahl) Quél. [= *Armillariella mellea* (Vahl: Fr.) Karst. = *Clitocybe mellea* Vahl], a parasite of prime importance on vines, fruit trees and numerous forest broad-leaved trees growing in many parts of the world. In Europe, its distribution can be called Atlantic-Mediterranean, since it is not found beyond Denmark to the north and increases in frequency in areas influenced by the Atlantic ocean and even more so in those influenced by the Mediterranean sea, whereas it is rare in Central and Eastern Europe under continental climates, though it was observed in Georgia and southern Russia. The fungus was also found in Syria and North Africa, and there are reports from the tropical Africa too (Kenya, Tanzania, Zaire). In North America, it is confined to the U.S.A., where is found on both the western (California) and the eastern side, while it becomes less frequent in the interior.

Its areale, that is fairly thermophilous, and its predilection for compact, fundamentally asphyxiating soils partly explain its limited incidence on genus Populus utilised in intensive plantations, which are characterised by a short cycle and in preference grow on loose and well-aired soils. Both in natural and natural-like forests its presence on over-mature plants, or on those weakened by adverse environmental conditions, is more common, although always sporadic. In Europe, Euramerican poplars close to maturity are sometimes attacked in some parts of France as well as in the centre and south of Italy, always on soils conditioned by water stagnation.

Although the epigeal symptoms are similar to those caused by R. necatrix (decline of sectors of the crown, stunted vegetation etc.), examination of the lower part of the trunk and of the roots next to the collar enables A. mellea to be identified from the presence of crimson, palmate, subcortical mycelial plates, often accompanied by typical flat black rhizomorphs (2-8 mm wide). The parasite can also climb up under bark tissues of the trunk to more than a metre from the collar, as well as penetrate the wood via the medullary rays causing fibrous decay. The typical fruiting bodies (basidiomata) of the fungus can form at the base of the most debilitated plants. They have a honey-yellow to brownish pileus (cap) (Ø = 3-10 cm), sprinkled with small transient scales that become fewer towards the edge, and with many lamellae (gills), whitish at first then reddish at maturity, decurrent along the stipe (stalk); this is fibrous, circular (Ø = 1-3 cm), 5-12 cm long and darker in its lower part, characterised by an armilla (ring) with yellowish stripes on its under surface.

The hyaline, ellipsoid basidiospores (7-10 × 5-7 μm) are of little account in the propagation of the fungus, which is mainly entrusted to root-to-root contacts with neighbouring plants and to the possibility of vegetating in the soil through the colonisation of organic debris. In the case of A. mellea, which is more pathogenic than saprophytic, rhizomorphs running in the soil have a limited growth capacity and hence are of little consequence as propagation and/or resistance organs; they are primarily active structures for the infection of roots not too distant from the starting substrate. Since it is a weakness parasite, albeit endowed with considerable pathogenic potential, A. mellea afflicts plants suffering from some form of stress. In poplar growing, this stress is generated by the excessive reciprocal competition for nutrients in crowded mature plantations and by recurrent leaf diseases.

As regards other Armillariae on poplars, in Europe the sole A. gallica Marx. et Romagn. [= A. bulbosa (Barla) Kile] was very sporadically reported in the United Kingdom and France, but it is generally much less pathogenic.

Some species have been found in association with dead or greatly debilitated plants in the north-eastern U.S.A., though the extent to which each of them is really parasite or saprophyte is still uncertain. A. ostoyae (Romagn.) Herink [= A. obscura (Schaeffer) Herink], which primarily attacks conifers, has been found on trees and suckers of P. tremuloides and P. grandidentata Michx., where it is thought to be an active root rot agent. A. gallica, probably as a secondary parasite or saprophyte, and A. sinapina Bérubé et Dessur., a species with greater pathogenicity, were also found on the aforesaid hosts. A. nabsnona Volk et Burdsall ranges from California to British Columbia and Alaska; it is a recently established species that was reported on P. trichocarpa T. et G.

DISEASES OF LEAVES AND YOUNG SHOOTS

Rusts Caused by Melampsora Spp.

Leaf parasite fungi belonging to genus Melampsora Cast., now recognised as the only member of the fam. Melampsoraceae (ord. Uredinales, class Teliomycetes, phylum Basidiomycota), are solely responsible for all poplar rusts.

Over the last twenty years, the leaf diseases known as rusts have been increasingly limiting poplar culture in countries such as Belgium, France and Italy, where it has long been established, and have hampered its further expansion in those where it has been recently introduced, such as Australia, New Zealand, southern Africa and Argentina. At the world level, they are one of the poplar's main enemies and as such cannot be overlooked by the researcher in his selection of clones and choice of control strategies, nor by the grower in the establishment and management of his plantations.

Symptoms and the damage caused-The presence of a rust agent is solely determined by the extent to which its host is genetically susceptible.

This description of rust symptoms mostly refer to intensively cultivated black and hybrid poplars belonging to the Aigeiros section but, with some variations, they are equally applicable to most rusts infesting poplars of the other sections.

First attacks in the plantation take place from early to late summer, depending on the geographic region, and a little earlier in the nursery, where its beginning is accelerated by high humidity and crowding of the saplings.

On susceptible plants, the first signs are minute roundish chlorotic spots with indistinct edges, randomly spread over the leaf abaxial side, which gradually become more evident and circumscribed while others continue to appear. After 2-3 days, the more developed spots give rise to small yellow to orange pustules (depending on the agent involved), that over the course of a few hours grow into irregular shapes after lacerating the leaf cuticle. These powdery formations, called uredinia, represent one of the anamorphs of the pathogen and are composed of masses of urediniospores, which are able to reinfect distant plants since they are readily dispersed by light breezes and, probably, also by the bodies of some insects. Very young and very old leaves are less susceptible than those of middle age, therefore uredinia are usually first noted at a certain distance from the apical leaflets, and then spread to the leaves close to them. As the growing season proceeds, repeated infections on the same plants result in thick uredinia over most leaves, which are thus sprinkled with a fine yellowish-orange «powder» that is the typical symptom of the disease.

In this stage, extensive portions of the leaf blade parch, owing to the confluence of the small necrotic areas around uredinia, or show conspicuous yellowing. This helps in sharpening the distress of the plant, which is also caused by excessive transpiration via the lacerated leaf cuticle and by a general weakness due to consumption of nutrients by the parasite. Precocious defoliation during the summer weakens the shoots and makes them unable to withstand the rigours of winter. It has also been shown that the marked decline exposes the hosts to attacks of weakness parasites in the following year, especially on the part of the bark pathogen Discosporium populeum.

In autumn, on the portions of the axial and/or abaxial side-depending on the species of Melampsora-of the senescent or already fallen leaves originally occupied by the uredinia, irregular blackish crusts (up to 1 mm), named telia, develop and remain in the litter throughout the winter. Telia, which secure the survival of the parasite till the next growing season, are seen as the teleomorph of rust agents. Such heavy levels of attacks are mostly encountered in nurseries, where the tips of the saplings may even die, and quantitative damage in terms of impaired growth may thus be converted into total loss of the production because it is not up to commercial standards. In plantations-except the thick mini-rotations for the production of biomass in semiarid areas (such as the Near East)-the disease rarely proceeds further than moderate defoliation, because the microclimatic conditions are less favourable and leaf formation slows down in the height of summer. Even so, the cumulative effect of attacks during several growing seasons may lead to a substantial loss of woody growth and predisposition to drought stress, due to insufficient root development.

Life cycle-The Melampsorae are obligate parasites that draw nutrients from live plant tissues in every stage of their life cycle. This includes several types of anamorphs and spores, each with a well-defined reproductive role, and also requires for its completion the passage of some stages on other hosts, named secondary hosts, which may be arboreal or herbaceous depending on the rust species and are not taxonomically related to Populus.

Briefly, a typical life cycle develops as follow. Telia, that form in the autumn on senescent and fallen leaves, are aggregations of elongated spores, arranged like a palisade and known as teliospores, which remain dormant during the winter. In spring, when suitable conditions of temperature and humidity take place, each teliospore emits a short germ tube on which basidiospores, derived from the conclusion of the sexual process, are formed. Basidiospores cannot reinfect the poplar, but need to pass to a secondary host and infect its green tissues. The first, not every evident, anamorph fructifications, called spermogonia or pycnia, appear about ten days later; their task is to produce tiny gamete-like propagules, called spermatia, in a sticky sugar-rich substance, and to capture, by means of receptive hyphae, spermatia of the opposite mating type brought by raindrops and some insects from other pycnia. The true fecundation takes place at the base of the pycnia and leads to the formation of a second type of anamorph fructifications, the aecia. These elongated, yellowish, powdery masses, that burst out of the cuticle of herbaceous tissues or of needles (in case the host is a conifer), are long chains of aeciospores, which are easily airborne and have a durable viability. Aeciospores are able to reinfect poplar, i.e. the primary host, through entering leaf stomata. Uredinia are then formed on the leaf abaxial side throughout the remainder of the growing season, owing to the production of at least two or three urediniospore generations. In autumn, the overwintering telia differentiate on senescent or fallen leaves.

Teliospores are of great biological significance, since they are the scene of the conclusion of the sexual process, i.e. the modifications of the genetic complement that will be transmitted to basidiospores and perpetuated through the next cycle.

In many areas both in and outside Europe, some poplar Melampsorae resume their production of urediniospores in the next vegetative season without passing to the secondary host, because they overwinter either as mycelia close to the dormant buds or, in small percentage and in places where the winter is not too cold, as urediniospores that remain attached to the dry fallen leaves. When designing a plantation, a grower must none the less have a sufficient knowledge of the secondary hosts and of their proximity to the land he intends to use, since their presence favours genetic recombination of the parasite, which can only take place if its cycle is completed, and hence the probable propagation of new pathogenic strains.

The pathogens-Many Melampsora species, whose taxonomy is the subject of current discussion, can infect Populus, though only a few are of phytopathological significance. Rusts on black and hybrid poplars belonging to the Aigeiros section (P. nigra, P. deltoides, P. × euramericana) are of prime concern on account of the economic damage they are now causing to intensive cultivations on the world scale as well as of the risk associated, even in the short term, with their proven genetic dynamism, which raises serious strategic problems for the selector of clones. Three species, with different degrees of aggressiveness against currently employed clones, are responsible: M. allii-populina Kleb., M. larici-populina Kleb. and M. medusae Thüm.

M. allii-populina is relatively thermophilous and covers a large areale that once extended from northern Africa to western Asia and much of south-central Europe, but now embraces southern Africa as well. It produces telia on the leaf abaxial side and uredinia usually dark orange. Even though it can perpetuate infections on poplar (paracycle), it completes its life cycle on herbaceous plants of the Allium L., Arum L. and Muscari Mill. genera. Two formae speciales were recently stated: f. sp. allii-populina, pathogenic on various Allium and Arum species, and f. sp. muscaridis-populina Vien., found solely on M. comosum (L.) Mill.

M. larici-populina is probably the most cosmopolitan, since it occurs wherever poplars are grown. Initially confined to Eurasia, it spread to South Africa, Australia and New Zealand in the 1970s and to Washington, Oregon and California over the last ten years; it was also reported in Argentina. In the latest years, it is responsible for repeated epidemics on European and Australian intensive cultivations of Aigeiros species, and has also been observed both in the U.S.A. and Europe on species from other sections, e.g. P. trichocarpa or P. × interamericana, which are widely planted in regions of the northern hemisphere under continental climates. In addition to producing its telia on the leaf adaxial side, it is distinguishable from M. allii-populina through some microscopic characters Its secondary hosts are some larch species, though in the U.S.A. it was recently found on Pinus ponderosa and P. contorta too; like M. allii-populina, it can perpetuate on poplar in the uredinial stage.

M. medusae, original of North America, is now widespread in Argentina, South Africa, India, Japan, south-eastern Australia and New Zealand too, whereas in Europe it is currently confined to Spain, Portugal and south-western France. By contrast with the aforesaid species, it needs to complete its life cycle on secondary hosts, these being almost solely North American species of Pinus, Larix, Pseudotsuga, Tsuga, Abies and Picea. Morphological, biological and epidemiological identities led to the synonymy of M. albertensis Arth. and, according to the latest interpretations, of M. abietis-canadensis (Farl.) C.A. Ludwig with M. medusae. In addition to Aigeiros poplars, where it is active as f. sp. deltoidae Shain, it is able to infect P. tremuloides and P. grandidentata (as f. sp. tremuloidae Shain) as well as several Tacamahaca poplars and intersectional hybrids (e.g. P. × interamericana). M. medusae produces its telia both on the leaf adaxial and abaxial side.

Mention may also be made of M. medusae-populina Spiers, a new species recently discovered on black poplars in New Zealand, with a middle morphology between M. medusae and M. larici-populina.

M. occidentalis Jacks., a species currently confined to the U.S.A. and Canada, is very morphologically and symptomatologically similar to M. larici-populina, but it has a much broader spectrum of secondary hosts. Its phyopathological importance, however, is negligible.

Leuce poplars are not often cultivated intensively and hence their rusts are of minor economic significance. In Europe, the main species found on Albidae and Trepidae are M. pinitorqua Rostr., M. larici-tremulae Kleb. and M. rostrupii Wagn. Some authors still refer to them as the M. populnea complex on account of their similar characters and the same primary hosts (P. alba, P.tremula, P. canescens).

In north-central China, M. rostrupii and M. magnusiana Wagn. are responsible for heavy nursery losses, especially at the expense of P. tomentosa Carr.

Relations with the host-Like all the rust agents, the Melampsorae are obligate parasites of living tissues; therefore, they do not kill the cells they colonise, but absorb some of their substances, particularly certain sugars and aminoacids. The parasitic association is very complex, and the extremely specific compatibility between the parasite and the host stems from reciprocal recognition on the part of the former's genetic sequences and their complementary sequences of the latter. This recognition underlies the existence of what are known as physiologic races, differing solely in their range of susceptible clones but morphologically identical, which are responsible for the recent increase in Europe of the severity and spread of attacks by M. larici-populina, now virulent on previously resistant clones. Five races have been reported in about ten years, while the presence of others is suspected, even in places such as the western U.S.A., where the parasite has only spread recently. M. allii-populina and M. medusae too seem involved by this phenomenon, whereas in this connection little is known of rust agents reported on the other poplar sections, one reason being their minor economic importance.

Control strategies-A balanced, effective protection against Melampsorae, as far as possible, must be primarily sought through quarantining and the search for resistant clones and, subordinately, through growing practices and a limited use of fungicides.

A close check must be kept on the international exchange of plant materials, since the recent spread of various rust agents to new regions has been partly due to ill-advised importations. Nevertheless, these provisions are not absolutely reliable, since independent dispersion of rust fungi over long distances is always possible.

Satisfactory short-term results can be obtained through the selection of resistant or tolerant clones capable of withstanding other adversities, adaptable to various environments and sufficiently productive. For example, many P. deltoides clones are resistant to M. larici-populina and M. allii-populina, whereas in P. × euramericana different reactions are showed according to the various genotypes. Unfortunately, in the long term the emergence of new physiologic races, able to attack clones previously shown to be resistant, may well stultify the selector's efforts (e.g. M. larici-populina race E3 to the detriment of "Luisa Avanzo" Euramerican clone).

Appropriate growing practices are thus indispensable, especially in nurseries and mini-rotations for the production of biomass. They range from eradication of the secondary hosts (sound practice for M. allii-populina) in areas near the plantation, to covering of propagule-bearing leaf residues with earth to a sufficient spacing of cuttings to reduce the excessive dampness of the microclimate.

In epidemic areas, the aforesaid practices can be integrated with fungicides treatments at fortnightly intervals from the appearance of the first uredinia, though not more than 2-3 times in a single growing season. Good experimental results were achieved with suspensions containing organic nitrogen compounds, benzimidazoles and ergosterol biosynthesis inhibitors; myclobutanyl, triadimenol, tebuconazole and cyproconazole are among the most effective active principles.

Biological control through the use of microorganisms naturally present on poplar leaves also furnished interesting experimental results. Cladosporium Link spp., Alternaria alternata (Fr.) Keissler, Sphaerellopsis filum (Biv.-Bern.: Fr.) Sutton etc. demonstrated a certain ability to inhibit the sporulation of rusts.

Symptoms and the damage caused-The Marssoninae are able to attack healthy plants. Symptoms vary in function of the parasite species and the host.

On Aigeiros poplars, M. brunnea, by contrast with the Melampsorae, is mainly active in plantations and preferably attacks young leaves and shoots still in the herbaceous state. Leaves starting from the stage of development just prior to extension and for about 10 days are particularly receptive, whereas a refractivity to infection hardens as the tissues mature; nevertheless, senescent leaves reacquire a certain degree of susceptibility, especially in late summer. In spring, brownish spots (Ø _ 1-2 mm), paler in the centre, appear on both sides of the leaf, scattered at first, but soon confluent owing to their thickening.

The surrounding leaf tissues turns yellow, then brown, thus leaves as a whole acquire a bronzed appearance that is typical of the disease and usually the prelude to their early falling off. On vascular bundles, petioles and other green parts of the plant (vegetative tips, current year branchlets, inflorescences and then infructescences), similar spots-though a little larger and elongated-now appear, within which the parasite produces a large number of conidia inside fruiting bodies known as acervular conidiomata, that are visible to the naked eye as whitish waxy lumps. As already stated, the main attacks occur in plantations where, starting from the lowest branches, the typical bronzing of the foliage is soon followed by a more or less marked early phylloptosis, though the apical portion of the foliage is spared. Appearance of the disease in the nursery is less common, especially in the first year.

M. tremulae, now more correctly transferred to M. brunnea f. sp. trepidae, is responsible for similar symptoms on P. tremula and P. tremuloides. Rather larger brownish spots (Ø = 2-3 mm) appear, usually on the leaf abaxial side, and then join together to cause large tissue necrosis. However, the ensuing phylloptosis is usually rather late and confined to the lower branches and the suckers.

The syndrome caused by M. populi takes the form of much larger roundish spots (Ø _ 4-5 mm), mainly on the leaf adaxial side, with an outline that is distinct on the Euramerican hybrids, frayed on P. nigra. These blotches extend over much of the leaf through confluence, whereas large lozenge-shaped lesions appear on the still green shoots. The attacks of this species are almost never of any epidemiological importance.

M. castagnei causes spots on P. alba that are similar in size to those induced by M. populi, though their edges are never frayed. These too are usually found on the leaf adaxial side, since their appearance is hindered on the abaxial side by its thick tomentum. With a shape that is often irregular, they are greyish-white in the centre and, at a certain stage, sprinkled with tiny waxy protuberances, i.e. the fruiting bodies of the parasite. On the petiole smaller spots elongated in the direction of its axis often appear, whereas traces of infections are rare on shoots. The disease causes marked yellowing, especially of the lower branches, followed by early phylloptosis.

Repeated attacks by M. brunnea on intensively cultivated sensitive Euramerican clones lead to gradual impairment of their annual wood production, and also make them predisposed to infections by weakness parasites, agents of bark necroses, and to the so-called "brown spots" coming from physiological stresses. Death of the trees belonging to the more susceptible clones is exceptionally reported.

The pathogens-According to the latest taxonomic revision, Populus is attacked by four Marssonina species, differing in their natural hosts and in some features of their conidia

M. brunnea (Ell. et Ev.) P. Magn. (= M. populicola Miura) is by far the most important in both phytopathological and economic terms. Of little consequence until the end of the 1950s and endemic in well-delimited areas of the U.S.A., it since spread first to Japan, then to Italy and most of the rest of Europe, where it caused ruinous epidemics on plantations still in full expansion, and subsequently to other Asian countries, New Zealand (1976) and Argentina (1990). Today it is present with varying degrees of incidence wherever poplars are grown, with the sole exception of Africa. Spiers (1984) recently distinguished two formae speciales with different hosts: f. sp. brunnea, which attacks P. nigra, P. deltoides and above all P. × euramericana, but not P. tremula and P. tremuloides; f. sp. trepidae Spiers (= M. tremulae Kleb.), which is only pathogenic on the last two species. Even so, f. sp. brunnea is also able to infect members of the Tacamahaca section (P. trichocarpa, P. simonii Carr.) and the intersectional hybrid P. × interamericana (north-western U.S.A.). According to what was stated for the anamorph, Spiers & Hopcroft (1998) for reasons of precedence transferred to Drepanopeziza tremulae Rimpau D. punctiformis Gremmen, the ascomycete that was previously described as the teleomorph of M. brunnea.

M. populi (Lib.) P. Magn. (= M. populi-nigrae Kleb.), whose teleomorph is D. populorum (Desm.) Höhn., is found throughout Europe and in Russia, China and North America, but is of limited phytopathological importance. It is more active on P. nigra (particularly on Lombardy poplar) and some P × euramericana clones little used for intensive cultivation than on P. deltoides and other hybrids; besides, it can infect P. tacamahaca and some Aigeiros × Tacamahaca hybrids. It differs from M. brunnea in the symptoms induced on the host and in having curved, pear-shaped conidia.

M. castagnei (Desm. et. Mont.) P. Magn. (= M. populi-albae Kleb.), with teleomorph D. populi-albae (Kleb.) Nannf., was widespread throughout Europe, the Middle East and America before being reported in Australia and New Zealand in the mid-1980s. By comparison with other countries, in New Zealand it has displayed a certain degree of aggressiveness and triggered by no means negligible epidemics on the "silver poplar" cultivar of P. alba, widely used locally for soil conservation. In addition to this species, which is virtually its only host in nature, it can infect some Albidae × Trepidae intrasectional hybrids as well as Leuce × Aigeiros and Leuce × Tacamahaca intersectional hybrids, but curiously it is not reported on Trepidae poplars.

M. balsamiferae Y. Hirats., the most recently described species, was only reported in 1984 on three herbarium specimens of P. balsamifera collected in Canada (Manitoba and Ontario). Its curved conidia with pointed tips quite distinguish it from the other three species and are more reminiscent of Neomarssoniella U. Braun.

Life cycle and relations with the host-The Marssoninae are leaf parasites like the rust agents but, unlike them, they kill and degrade the tissues they colonise through a proper enzymatic action. This means that their relationship with the host does not reach the levels of specificity displayed by the Melampsorae, since the metabolic «recognitions» with the host living cells and the close relations between portions of its genetic material and those of the parasite are lacking. This behaviour may also explain why no physiological races have yet been identified within Marssonina species, although proper studies were made.

Acervular conidiomata, representing the anamorph of the fungus, are differentiated in the centre of the spots from 7 to 10 days after inoculation on susceptible clones and in optimum atmospheric conditions, i.e. rain and/or occult precipitations and temperatures between 17 °C and 22 °C. Those of M. brunnea and M. castagnei are always intraepidermal, those of M. populi can also be subcuticular. Each conidioma (Ø = 200-400 μm) is composed of a rather slack hyphal matrix supporting a layer of short conidiophores, which produce two-celled hyaline conidia whose shape vary from one species to another, and whose common feature is the proximal cell (corresponding to the insertion on the conidiophore) smaller than the distal one. After rupture of the overlying epidermal wall and/or the cuticle of the host, the conidia are released in bulk in a mucilaginous substance and dispersed mainly by rain or dew, which dilute them into an aqueous suspension, and by the wind to considerable distances.

On coming into contact with susceptible green tissues and favoured by a film of water, the conidia germinate. The germ tubes, which may originate from either cell, are able to directly perforate the epidermis and to colonise the mesophyll cells without needing to pass through the stomata.

If the weather is favourable, during a vegetative season the succession of some peaks of conidium production result in one infection cycle after another. When autumn approaches, the usual conidia are accompanied by ellipsoid one-celled microconidia that probably serve as spermatia for differentiation of the teleomorph. The non-finding of the latter in New Zealand, even though M. brunnea and M. castagnei made remarkable inroads in recent years and despite the presence of their microconidia, suggests that each fungus is heterothallic and that in this country one mating type is absent.

Formation of the teleomorph fruiting bodies begins in autumn on the fallen leaves, when portions of mycelium clump together in the mesophyll to build the ascomata. At the start of the following spring, the ascomata, now in an advanced stage of maturation, burst from the tegument in response to progressive pressure to take on the typical lenticular shape of apothecial ascomata with no stalk (measuring 100-300 μm depending on the species). The fertile layer, exposed to the outside in a stage of full maturation, is composed of paraphyses and asci, which are elongated sack-like cells each containing eight hyaline, one-celled, ellipsoid ascospores (10-16 × 3-9 μm depending on the species).

By contrast with the conidia, the ascospores are chiefly dispersed by currents of air and are responsible for primary infections when the vegetative season recommences. Their real epidemiological importance, however, is determined by both the geographic region and its climate: for example, in north-central Europe they seem usually determinant, whereas in southern Europe their role is totally secondary by comparison with the copious attacks resulting from conidia, which are more favoured by the milder climate and thus already in progress before the ascospores are dispersed. This pattern stems from the fact that the parasite is able to overwinter in the forms of stromata (i.e. compact masses of vegetative hyphae), particularly in the cortical tissues of already infected branchlets, which produce new conidia infecting the sprouting leaves when growth recommences.

Control strategies-Being stated that the exchange of biological material between one country and another must be strictly supervised by appropriate quarantining, to prevent or at least delay the further spread of the parasite, as in the case of the agents of other specific diseases resistant clones must be selected and employed.

Genetic improvement and the selection of clones have so far been favoured by the continuing absence in the Marssoninae of physiologic races. This type of control, however, must also take account of the resistance to other diseases and the adaptability to the environment of genetic material previously obtained in the absence of Marssonina. For example, the resistance of several P. deltoides clones to M. brunnea is opposed by their susceptibility to the poplar mosaic virus and their vulnerability to wind damage; that of some highly productive Euramerican selections by their predisposition to physiological stress and hence to the appearance of "brown spots", or by their susceptibility to Pollaccia elegans and to Melampsorae. The same can be said for some P. × interamericana selections, whereas other new clones, some of them intersectional too, seem to offer better prospects of success.

Among indirect measures, though almost never resolutive, correct cultivation practices are advisable, especially during the parasite's winter quiescence, to reduce the mass of inoculum and hence the heaviness of the subsequent attacks. The maturing ascomata and stromata hibernating in the fallen leaves can be dug in, those hibernating in the bark of the branchlets can be destroyed by pruning and burning. Resistance can also be raised, within certain limits, by the application of potassium and nitrogenous fertilisers.

Increasing awareness of ecological repercussions of the use of chemicals has placed a brake on this form of control. Nevertheless, In European and North American plantations susceptible to M. brunnea and in Oceanian plantations susceptible to M. castagnei as well, it may be necessary to employ fungicides systematically. Attention is primarily directed to preventive treatments with dithiocarbamates, i.e. maneb or mancozeb (200 g/hL a.i.).

Curative treatments, however, are also possible using fungicides with systemic action, such as triforine (60 g/hL a.i.), dodine (70 g/hL a.i.) and triadimenol (10 g/hL a.i.). To be effective, these treatments must be performed in spring starting from the leaf sprouting time (no more than two or three treatments at monthly intervals), and applied with either land-based nebulizers or atomisers mounted on fixed-wing aircraft or helicopters, which ensure the maximum promptness and effectiveness on large areas.

Leaf Scab and Blight Caused by Pollaccia Spp.

The form-genus Pollaccia Bald. et Cif.-taxonomically revised on several occasions in recent years-includes the anamorphs of parasites on poplar leaves and shoots whose teleomorphs, only occasionally observed in nature, are ascribed to Venturia Sacc. (fam. Venturiaceae, ord. Dothideales, phylum Ascomycota).

The main pathogens are P. elegans, the subject of particular attention owing to its renewed aggressiveness on black and euramerican poplars cultivated in some European and North American areas, and P. radiosa, which attacks many Leuce poplars in several continents. This section is also infected by P. borealis and P. populi-albae on a more local scale. Finally, P. mandshurica is currently endemic in China on some members of various sections. Symptoms and the damage caused-Healthy plants from susceptible clones, in cool stands, are attacked by these pathogens at the beginning of the spring.

P. elegans is primarily active in the plantation against Aigeiros poplars; its targets are the first leaves sprouting from dormant buds and the shoots of the current year. On leaves, it causes brown-yellowish lesions irregular in shape, though mostly rhombic, that usually are placed along the main bundles. Confluence of these lesions results in extensive necrosis and turns the leaf blackish. A few days after the appearance of the first symptoms, an olive bloom, due to the mass production of conidia, thicken on the dead leaf tissues. Almost at the same time, the still herbaceous shoots display depressions and necroses, often located at their base owing to the passage of the parasite through petioles. The distal tissues gradually lose their turgor, bend into a hook shape, dry out and eventually split. In more resistant clones, reddenings and swellings form around the infection sites and block any further colonisation. Widespread and intensive attacks can impair all the leafage put out at the beginning of the vegetative season, but the host is able to activate the buds below the necrotised sprouts and thus to form a new crown, that is not usually stricken by new infections. Woody growth is only impaired in the event of repeated severe attacks during some years.

The symptoms caused by P. radiosa and P. populi-albae on Leuce poplars are much the same, though the leaf lesions are more numerous, less extensive and irregularly roundish. Their colour ranges from olive to brown with a darker border, and changes to reddish-brown in the centre at the time of sporulation.

P. borealis has been recently been reported on Populus tremuloides. By contrast with those of P. radiosa, the lesions it induces are brown-purple, limited in size (Ø _ 0.5 cm) but confluent, surrounded by a dark brown ring that is more evident on the leaf abaxial side, the only place where sporulation occur. The leaves of some clones are pierced by holes which are distinctly smaller than ordinary necroses and have a thickened inner rim. They can readily be mistaken for the work of insects. By contrast with the attacks of other species, petioles and shoots show no sign of infection.

P. mandshurica is the agent of the syndrome known in China as "grey spot disease". Primarily serious in the nursery and on young trees up to 5 or 6 years old in the plantation, it causes lesions similar to those provoked by P. radiosa, except that their inner part is silver grey and bounded by a dark brown edge.

The pathogens-The Pollacciae found on poplars differ in their elective hosts, and also in some morphological and morphometric features of their fruiting bodies and propagules Many disputes have arisen concerning the species which prefer the Leuce section.

P. elegans Serv. [teleomorph: Venturia populina (Vuill.) Fabr.] is widespread in Eurasia on Aigeiros and Tacamahaca poplars, in North America only on Tacamahaca (P. trichocarpa and P. balsamifera in the North West Territories and in Alaska) and P. trichocarpa × P. deltoides hybrids. It is currently present in a latent form in Europe, where it causes sporadic infections in very damp springs, but retains a core of higher aggressiveness in the Alps and subalpine regions (at 200-800 m above sea level) and in the eastern Po Valley. In this zone, the use of Euramerican clones selected from "Canadian" genotypes, in its turn imposed by repeated attacks of Marssonina brunnea on the established clones, has been responsible for a certain recrudescence in recent years (1990-1993). P. radiosa (Lib.) Bald. et Cif. (teleomorph: V. tremulae Aderh.) was divided by Morelet, as accepted here, into two varieties that differ in their micromorphology and (partially) in their hosts, but that induce virtually identical symptoms:

- var. radiosa (teleomorph: V. tremulae Aderh. var. tremulae), which is spread in Eurasia and North America on Populus tremula, P. tremuloides, P. canescens, P. alba and various intrasectional hybrids;

- var. letifera (Peck in Sacc.) Morelet (= P. americana Ondrej), whose teleomorph is V. tremulae Aderh. var. grandidentatae Morelet, spread in Canada and U.S.A. on Populus grandidentata, P. tremuloides, P. alba and various intrasectional hybrids.

P. radiosa was once thought to be the anamorph of V. macularis (Fr.: Fr.) Müller et v. Arx, but the latter is actually a saprophyte fungus (like V. viennotii Morelet) of no phytopathological importance, whose anamorph is for the present unknown, found on dead leaves and petioles of Populus grandidentata, P. tremuloides, P. tremula and P. canescens.

P. populi-albae (Morelet) Rulamort [= P. ramulosa (Desm.) Ondrej], whose teleomorph is V. tremulae Aderh. var. populi-albae Morelet, attacks P. alba only both in Europe and North Africa.

P. borealis Funk differs from P. radiosa in the appearance and course of the lesions it causes on leaves, as already mentioned, and in some micromorphological characters. Its areale includes parts of British Columbia far from the Pacific coast, the North West Territories and Yukon, where it is responsible for substantial epidemics on P. tremuloides. The teleomorph V. borealis Funk was obtained on artificial medium, but has not yet been observed in nature.

P. mandshurica Morelet (teleomorph: V. mandshurica Morelet) is endemic in north-eastern China in the region once called Manchuria, where it induces a syndrome first recognised in the early 1960s, though the fungus was not described and identified as the pathogen until the 1990s (owing to constant confusion with the bark parasite Coryneum populinum Bres. and with Mycosphaerella mandshurica Miura in the case of its anamorph and teleomorph respectively). Populus simonii, P. nigra, P. davidiana and various P. simonii × P. nigra hybrids are susceptible to its infections, whereas P. alba, P. trichocarpa, P. deltoides and P. × euramericana hybrids are resistant.

Life cycle and relations with the host-The Pollacciae are favoured by falls in temperature following the vegetative restarting of buds in the presence of moderate precipitations and high humidity, a common situation in early spring, both because they are fungi that vegetate well at quite low temperatures (the optimum range is 15-20 °C) and because the susceptibility of the shoots is enhanced in these atmospheric conditions. P. mandshurica alone mainly infects during the summer; in north-eastern China, in fact, concentration of the rainfall in temperate summers favours later attacks.

Primary infections are carried out by ascospores. They are two-celled, yellow-greenish (22-25 × 10-14 μm), released from typical pseudothecial ascomata that, in the case of V. populina, are differentiated in particular on the dried tissues of necrotised shoots bordering the still vital parts, and also on infected leaves that have fallen to the ground in the case of V. tremulae and V. mandshurica. At least in European poplar districts, infections on the part of the conidia released from stromata formed by the hibernating mycelium in the necrotised tissues, mainly inside the shoots killed the previous year, appears to be more important for the initial propagation of the disease, whereas the role of ascospores seems less important on account of their late maturation.

Subsequent infections are performed by conidia released from new typical cushion-like structures known as sporodochial conidiomata, which are composed of short hyphae and differentiated on the leaf necroses.

They continue to originate new infections as long as the favourable weather lasts (some workers maintain that there may be as many as five P. radiosa generations, each lasting two weeks, throughout the vegetative season), but usually their production ceases during the summer, thus the disease regresses. Resumptions in the autumn, when the optimum temperature and humidity values are re-established, are completely a matter of chance.

As in the case of the Marssoninae, no populations with different pathogenicity have yet been observed within the single Pollaccia species.

Control strategies-Particular attention must be directed to the selection and use of resistant clones. This has already led to lasting results and offered good incremental and technological parameters of the wood. In north-central Europe, for example, the preference for interamerican hybrids and P. deltoides clones as opposed to the euramericans limited the spread of P. elegans, despite the ideal climate. Care, however, must be taken to ensure that the clones selected are equally resistant to pathogens such as the Melampsorae, the Marssoninae, Discosporium populeum, Xanthomonas populi that are currently a major threat to poplar growing. Now that so much is known about the ease with which diseases can spread, the misfortunes that might arise from the imposition of a single resistance would be less excusable than in the past.

Cultivation practices able to reduce the incidence of the disease by removing the overwintering inoculum (pseudothecial ascomata and conidiogenous stromata), such as destruction of the infected shoots or turning over the soil to bury the fallen leaves, are costly measures and not always economically justifiable.

Systemic and contact fungicides too have failed to provide satisfactory practical results, not because their active ingredients are ineffective, but because treatments are rendered difficult by uncertain prediction of the time when the conditions favouring an attack are likely to occur. They are also very expensive and impose a heavy environmental burden if repeated.

Other Leaf Diseases

Yellow blister of leaves and amenta caused by Taphrina spp.-The syndromes attributable to some members of Taphrina Fr. (fam. Taphrinaceae, ord. Taphrinales, phylum Ascomycota), while encountered worldwide, are rarely the cause of substantial economic damage to poplars and are primarily known for their unmistakable characteristics.

The Taphrinae are obligate parasites of living tissues, hence do not kill the mesophyll cells or other green tissues before establishing the parasitic relationship, but they do not display a specificity comparable with that of the Melampsorae. To reproduce, they require a temperature of not more than 15-20 °C and high atmospheric humidity, thus their attacks mainly occur in spring.

The Taphrinae on leaves, while also found on adult poplars, are only a cause for concern in the nursery, where they take advantage of the closeness of the saplings and the particular microclimate to which this gives rise.

The most important species is T. populina (Fr.: Fr.) Fr. [= T. aurea (Pers.) Fr.], widespread in Eurasia, southern Africa, North America and South America (Argentina, Chile), also reported on willows. In the early days of poplar cultivation, the fungus had a certain impact in some European zones (e.g. in northern Italy) on black Eurasian poplars not yet selected for their disease resistance. It was soon brought back within latency level, however, through improved cultivation practices and empirical exclusion of the phenotypes that proved most sensitive. It is still quite common, especially in east-central Europe. The Aigeiros section is primarily affected (P. nigra and Euramerican hybrids, some of whose clones are anyhow poorly susceptible, such as "Robusta", or resistant, such as "I-214") and the Tacamahaca poplars are also attacked, though to a lesser extent.

Infections results in the appearance of blisters on the leaf blade, concave towards the abaxial side, isolated at first then confluent, up to 3 cm in diameter. The leaf adaxial side remains green for a long time, whereas the areas of the abaxial side corresponding to the concavities, derived from deformations of the mesophyll connected with hormonal disorders, turn bright yellow and then orange. This colour change is consequent to the reproduction stage of the parasite, provided by the formation of bare cylindrical asci (50-112 × 15-40 μm), arranged in a layer outside the leaf and supported by basal cells differentiated in the host's epidermal cells. Even before they mature, these asci contain droplets of a golden lipid substance whose mass gives its colour to the underlying areas of the leaf blade. The one-celled ascospores (Ø = 4-6 μm) are short-lived and generate many blastospores (Ø = 2-3 μm) by budding inside the asci, which eventually break open. The release of blastospores may be followed either by infections through development of subcuticular mycelium, or by direct production of other blastospores, which are then dispersed in the environment by the rain and the wind. At a late stage of the disease, large portions of the leaf necrotise and dry out, and holes are formed at the blisters. The fungus overwinters both inside the preformed leaflets of poplar buds in a mycelial state, and outside them as adherent blastospores with a thickened cell wall.

Another more localised leaf parasite is T. populi-salicis Mix, which is found in North America on willows and poplars (P. nigra, P. trichocarpa).

T. johansonii Sadeb. and T. rhizophora Johans. attack amenta at the ovaries and cause large, bright yellow swellings (hypertrophies). The first has been reported in Europe (mainly east-central Europe), Japan and North America on various members of Leuce and Aigeiros sections, the second only on P. alba in east-central Europe and Australia. The two species are distinguished by the length of their asci (more than 120 μm long in T. rhizophora).

Powdery mildews-For this group of diseases are responsible various members of the fam. Erysiphaceae (ord. Erysiphales, phylum Ascomycota). By contrast with forest nurseries and fruit-growing, they almost never cause serious damage in poplar cultivation.

The best known agent is Uncinula adunca (Wallr.: Fr.) Lév. [= U. salicis (DC) Wint.], widespread in Eurasia and North America on many Leuce, Aigeiros and Tacamahaca species, and also established on willows. The other agents reported on poplar are:

- U. populi Sharma, recently described on P. nigra in India;
- Phyllactinia guttata (Fr.) Lév. [= P. suffulta (Reb.) Sacc. = P. corylea (Pers.) Karst.], a cosmopolitan species found on numerous broad-leaved trees, but only reported on Euramerican poplars in southern Asia (Pakistan, India, China, Korea);
- Erysiphe horridula (Wallr.) Lév., observed in 1960 in Italian nurseries of P. nigra.

The powdery mildews are obligate leaf parasites like the Taphrinae, but are much less dependent on humidity and their attacks are thus more frequent in summer. A white mat of mycelium appears on the leaf adaxial side (and also the abaxial one in the case of P. guttata) which produces the typical conidia of the anamorph, responsible for the secondary infections, throughout the vegetative season. Leaves gradually curl up because the parasitised cells in the upper epidermis stop dividing, whereas those in the lower one continue to grow and divide. On the approach of the adverse season, globose corpuscles, known as cleistothecial ascomata, are seen for the first time on the white powdery mat. Yellow at first, then reddish and finally black as their maturation proceeds, they constitute the teleomorph of the parasite and contain many asci (in the case of species active on poplars). In the temperate regions, they are the form that overwinters on fallen leaves, while they also ensure the fungus dissemination by means of variously shaped external appendages which have a specific function of active transport (e.g. for P. guttata) or of attachment to the host surface (e.g. for U. adunca and U. populi).

The current incidence of the Erysiphaceae in poplar growing is too low to justify a special genetic selection of resistant clones or fungicide treatments.

Leaf blotch caused by Septotinia podophyllina-This pathogen (fam. Sclerotiniaceae, ord. Leotiales, phylum Ascomycota) was often referred to in the past as a poplar-specific species called S. populiperda Waterman et Cash, but later it was shown to be substantially coincident with the type-species S. podophyllina Whetzel, which has precedence and is also found on Podophyllum spp. and Prunus serotina Ehrh. It is thought to originate from North America, and has since been reported in some European countries (France, Holland, Jugoslavia, ex-Czechoslovakia, Russia) as well as in Korea and Japan. In the 1950s-1970s, it was the subject of a certain interest on account of its sporadic and occasionally serious recurrence in nurseries, where it damaged the production of cuttings (mainly in France), and in young plantations in the north-central U.S.A. There have been no subsequent reports about its activity.

The parasite is able to infect poplars of the three main sections (Leuce, Aigeiros, Tacamahaca), though with differences of clonal susceptibility. The disease appears in spring on young leaves as grey-brown spots, often placed on their edges or near insect punctures, that gradually expand following the mycelial growth when the humidity is high. Eventually each blotch reaches a remarkable size (up to 5 mm in diameter) and displays a distinct though irregular border, surrounding concentric light and dark bands corresponding to growth and static, low-humidity periods respectively, when leaf tissues react with a local accumulation of tannic compounds. On these blotches, which are readily visible on the leaf adaxial side, groups of whitish sporodochial conidiomata (Ø = 120-250 μm) produce fusiform, one-to five-celled hyaline conidia (18-45 × 5-8 μm) ascribed to the anamorph Septotis podophyllina (Ell. et Ev.) v. Arx. During the vegetative season, the blotches can converge and kill large portions of the leaf blade, causing precocious defoliation in serious infections. The parasite can also reach shoots via the petioles and induce dark, sometimes encircling lesions.

The teleomorph appears in winter, on fallen leaves, in the form of black circular or elongated sclerotia (3-5 × 1-2 mm), on which arise groups of cup-shaped apothecial ascomata (Ø = 2-7 mm) with long stipes and a greyish hymenium. Overwintering is ensured both by the hyaline ovoid ascospores (10-13 × 4-5 μm) and by conidia supported by sclerotia or produced in the lesions on shoots. Nevertheless, the real weight of these components in primary infections is a controversial question.

Leaf blight caused by Linospora spp.-Two species of Linospora Fuckel (fam. Valsaceae, ord. Diaporthales, phylum Ascomycota) are of a certain importance on poplar:

- L. ceuthocarpa (Fr.) Munk ex Morelet (= L. tremulae Morth.), spread in Eurasia, but also reported several times in the east-central U.S.A., agent of mild epidemics in France in the 1970s, and now with a purely secondary incidence;
- L. tetraspora Thompson, a northern species spread throughout Canada and increasingly in recent years, though still sporadically, in the northern U.S.A. too.

Besides having different areales, the two pathogens also have different hosts. L. tetraspora only attacks some Tacamahaca and Aigeiros American poplars (primarily P. balsamifera, followed by P. deltoides and P. trichocarpa × P. deltoides), whereas L. ceuthocarpa mainly invades Leuce species (P. alba, P. tremula, P. tremuloides, P. grandidentata), though it is occasionally found in nature on P. deltoides, P. × euramericana, P. trichocarpa and on intersectional hybrids (P. deltoides × P. trichocarpa, P. tacamahaca × P. deltoides).

At the height of summer, infections of Leuce poplars by L. ceuthocarpa cause grey-violet leaf spots visible on both sides, each measuring a few mm^2. Their further progression depends on the rainfall during the season: if this keep quite high, the spots expand rapidly at the end of August and the beginning of September in west-central Europe. They gradually become whitish in the centre, but remain dark on the edges, while their shape becomes circular or elliptical with wavy borders, often elongating at a main vascular bundle. There are rarely more than two or three blots per leaf, though the size they reach (as many as several cm^2) means that much of the same is involved, while its unaffected parts turn yellow. Brownish spots (1-2 mm) are sometimes observed on the petioles as well. Intensive attacks are rare; their immediate effect is early defoliation.

Black subcuticular acervular conidiomata (Ø = 100-350 μm) are soon differentiated in the whitish centre of the blots. When these mature, they lacerate the cuticle and release, in mucilaginous little masses, fusiform or sometimes curved hyaline conidia (7-24 × 2-3 μm), usually one-celled, often bound together by short lateral anastomoses. This is the anamorph, indicated as Titaeosporina tremulae (Lib.) v. Luik [= Gloeosporium tremulae (Lib.) Pass. = G. populi-albae Desm.]. The teleomorph appears at the end of the spring on dead leaves from the previous year, in the form of ovoid perithecial ascomata (average 400 × 140 μm) inserted between two parallel pseudostromatic laminae, visible to the naked eye as black, jagged notches, from which only the long necks of the ascomata protrude. The yellowish, filiform ascospores (100-130 × 1.9-2.7 μm), subdivided by 2 to 5 septa, are arranged in the asci in coiled bundles and are responsible for the primary infections.

Very similar symptoms are induced by L. tetraspora, with single blots that sometimes involve the entire leaf. Its ascomata, on the other hand, are smaller (175-270 × 110-175 μm) than those of the other species, and its ascospores are longer (175-225 × 2.5-3 μm) and have more septa (6-8).

Lastly, mention may be made of a number of poplar leaf parasites whose attacks are sometimes the cause of concern on a local scale or in particular contexts:

- Ciborinia whetzelii (Seaver) Seaver (fam. Sclerotiniaceae, ord. Leotiales, phylum Ascomycota), agent of the ink-spot disease mainly in young stands of P. tremuloides, sporadic throughout Canada and the northern U.S.A., which is so-called owing to the conspicuous black sclerotia formed on leaves; attacks on P. grandidentata, P. balsamifera and P. deltoides are rare;
- Glomerella cingulata (Ston.) Spauld. et H. Schrenk (fam. Phyllachoraceae, ord. Phyllachorales, phylum Ascomycota), with anamorph Colletotrichum gloeosporioides (Penz.) Penz. et Sacc. in Penz. (= Gloeosporium fructigenum Berk.), a polyphagous species responsible on poplar for moderate leaf and shoot blights, mainly in plantations of north-western America and Indian nurseries, reported in France as agent of brown necroses on one-year branchlets;
- Sphaceloma populi (Sacc.) Jenkins (= Hadotrichum populi Sacc.), the anamorph of Elsinoë populi Jenkins (fam. Elsinoaceae, ord. Dothideales, phylum Ascomycota), an agent of anthracnoses on Aigeiros and

Tacamahaca poplars in conditions of high humidity, mainly active in India but also reported in Europe, Japan and Argentina;

- Alternaria alternata (Fr.) Keissler (= A. tenuis Nees), a cosmopolitan mitosporic fungus very common on poplar as a component of its epiphytic mycoflora, which occasionally occurs as the agent of leaf blight in hot and very humid stands (especially in India and China);
- some species of Phyllosticta Pers. (Mitosporic Fungi), reported in Europe (e.g. P. populina Sacc.), Argentina, southern Australia, Japan and (as P. adjuncta Bub. et Serebr.) above all in India, which cause leaf spots on poplars rendered more vulnerable by particular growing conditions (e.g. very humid nurseries) or other diseases;
- two species of Phoma Sacc. (Mitosporic Fungi): Phoma exigua Desm., reported in New Zealand as a leaf blotch agent on one-year-old seedlings of P. alba, and Phoma macrostoma Mont., a polyphagous parasite that in India is sporadically responsible for leaf spots in nurseries and young plantations of P. deltoides;
- Phaeoramularia maculicola (Rom. et Sacc.) Sutton (= Torula maculicola Rom. et Sacc. = Cladosporium subsessile Ell. et Barth.), a mitosporic fungus which is agent of leaf spots on Aigeiros, Tacamahaca and above all Leuce poplars in North America and Scandinavia, as well as India, where its attacks have become increasingly frequent during the last twenty years;
- Cladosporium humile J.J. Davis, a mitosporic fungus responsible for severe phylloptoses in nurseries and plantations in various regions of India, especially on the indigenous P. ciliata, on which it is a pathogen of primary importance;
- Cercospora populina (Ell. et Ev.) Deighton, a mitosporic fungus that is found in India, where it gives rise to leaf blotch epidemics of a certain importance both in nurseries and plantations of P. deltoides, causing precocious defoliation in the older ones;
- Rhizoctonia solani Kühn., the anamorph of Thanatephorus cucumeris (Frank) Donk (fam. Ceratobasidiaceae, ord. Ceratobasidiales, cl. Basidiomycetes), an extremely polyphagous species with a wide variety of manifestations that on poplar causes a syndrome, known as leaf web blight, which reaches a high incidence in nurseries and young plantations in various regions of India subjected to heavy monsoon rains, thus requiring special cultural practices and fungicide treatments; its main symptom is a web of hyphae that is well visible on infected leaves like a cobweb and often stretches from one leaf to another.;

- Drechslera maydis Subram. et Jain [= Bipolaris maydis Shoem.], the anamorph of Cochliobolus heterostrophus (Drechsler) Drechsler (fam. Pleosporaceae, ord. Dothideales, phylum Ascomycota), long known as an important leaf parasite on maize, recently observed in India (isolates designated "T race") on some clones-mainly male of Texan origin-of P. deltoides, where it is responsible for a leaf blight of very high incidence; on the contrary, tests showed that P. nigra and P. ciliata are totally resistant.

According to the most recent taxonomic revision of Melampsora species found on poplars, which here is followed in part, the collective species M. populnea (Pers.: Pers.) Karst. [= M. populina (Pers.) Lév.] should include the former M. larici-tremulae, M. magnusiana, M. pinitorqua and M. rostrupii, together with M. allii-populina, as formae speciales: f. sp. laricis (Hart.) Boer. et Verh., f. sp. magnusiana (Wagn.) Bagyanarayana, f. sp. pinitorqua (Hart.) Boer. et Verh., f. sp. rostrupii (Wagn.: Kleb.) Boer. et Verh. and f. sp. allii-populina (Kleb.) Bagyanarayana respectively. This was suggested because these taxa «do not differ essentially in morphology, but only in their aecial hosts» (ibidem).

5

Mycology and Plant Pathology

MYCOLOGY

Fungi have the several features that distinguish them from other organisms. They have a filamentous branching system of cells, with apical growth, lateral branching and a heterotrophic nutrition. They are characterized by a life cycle that begins with germination from a spore or resting structure, followed by a period of growth as a substrate is exploited to produce biomass. Finally there is a period of sporulation, where propagules are formed that can be disseminated from the parent mycelium. Schematic of a fungal life cycle:

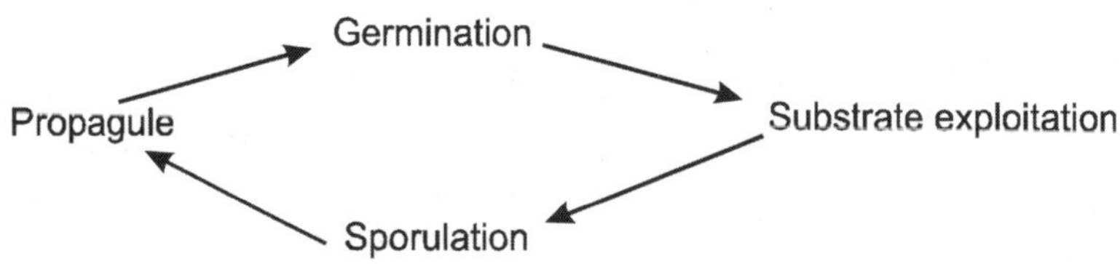

They are vital to the biosphere for many reasons, not least for their decomposing activities on dead substrates that ensure the release of nutrients like carbon, minerals and nitrogen back into the biosphere.

TAXONOMY

The origin of the fungi appears to be very ancient. They first appear in the fossil record coincidentally with the appearance of the land plants. This was in the Devonian period around 400 million years ago. The early microscopists found that fungi were simple to study, and they first observed division of cells by watching yeasts under the microscope.

They were one of the earliest groups to be classified, and the earliest scheme was drawn up in 1588. Early biological science was much obsessed with categorizing things and researchers constructed taxonomic systems based solely on structure. However, for the fungi, this created a polyphyletic group that contained microorganisms that had very different ancestors. Taxonomic thinking today now prefers a monophyletic classification, where all groups within a phylum are descendants of one ancestor. Therefore current taxonomy relies on studying characters that are features or attributes of an individual organism that can be used to compare it with another organism. These features can be morphological, anatomical, ultra-structural, biochemical or based on sequences of nucleic acids.

CHARACTERISTICS: BIOLOGY, ECOLOGY, USES, CULTIVATION

Mycology is the study of fungi and mycologists are the people who carry out these studies. New research methods have substantially increased knowledge about the fundamental nature of fungi. Much of this research has focused on fungi that cause plant diseases. Research on edible fungi has concentrated on a small group of species that are commercially cultivated. Wild edible fungi have, until recently, been relatively ignored by science, though amateur mycologists often documented species they found in field studies, mostly in Europe or countries in which Europeans have settled.

There has always, however, been a keen interest in a small group of valuable wild edible fungi that cannot be cultivated. These include the truffles (*Tuber* spp.), *matsutake* (*Tricholoma* spp.) and *porcini* or *cèpes* (*Boletus edulis*). Their biology and ecology have been studied in some detail – a marked contrast to the many other wild edible fungi used around the world. The consequence of this neglect is that wild edible species used in developing countries are poorly known. Some information is available from studies of close relatives in temperate regions. *Russula* and *Lactarius* occur around the world, for example, and knowledge of species in Europe can be applied with some caution and caveats to African species. The main problem is naming and recognizing species. Genera and species concepts were originally based on the narrower range of diversity found in temperate regions and these may require fundamental reappraisal as tropical species become better known. This chapter provides a brief introduction to the larger fungi (macrofungi), with special reference to those that are edible. The use of specialist terms has been avoided where simpler alternatives are available. Field guides contain useful glossaries and there are an increasing number of Web sites that help in understanding technical terms. The *Dictionary of the fungi* is a regularly updated text with details about all fungal genera and other information on mycology.

WHAT ARE FUNGI?

Fungi are a distinct group of organisms more closely related to animals than plants. At present fungi are divided into three separate and distinct kingdoms based on an expanded knowledge of their biochemistry and genetic makeup established especially over the last 30 or so years. It is wrong and misleading to refer to fungi as "plants without chlorophyll" (FAO, 1998a).

Despite fundamental differences, fungi are often classified as plants. Understanding the taxonomic status of fungi has little apparent significance to people collecting and selling wild edible fungi, but it is of critical importance in establishing a sound and robust classification system. This ensures that when two people use the same species name they know that they are referring to the same (edible) fungus. The classification of fungi with plants has inadvertent practical consequences. It is not always clear whether ethnobotanical studies include wild fungi, as is the case with a study from Turkey. Ethnomycology is the correct term that indicates fungi are involved. On a similar track, flora refers only to plants. The equivalent term for fungi is mycota. These fungal terms may be unfamiliar but their use helps to identify published information on wild edible fungi clearly that may otherwise be ignored or missed.

Structure and Feeding

Fungi come in many shapes, sizes and colours. Macrofungus (plural: macrofungi) is a general category used for species that have a visible (to the unaided eye) structure that produces spores, such as a mushroom or truffle. These visible structures are generically referred to as "fruiting bodies".

Fungi consist of fine threads known as hyphae, which together form a mycelium, as in the mould growing on a piece of fruit or bread. The cap of a mushroom or a bracket fungus also consists of hyphae, densely packed together to form the fruiting body.

Specialized hyphae produce spores that are dispersed in a number of ways. They can be viewed en masse by placing the cap of a mushroom on a piece of white paper and covering it with a glass. The colour, form and way in which spores develop help to identify the fungus. Wild edible fungi are often referred to generically as wild edible "mushrooms". This can be confusing for a number of reasons: edible species have different forms, some with gills and some with pores, some with stems and some without. This book prefers the broader term wild edible fungi to reflect the diversity of forms and also to distinguish them clearly from cultivated mushrooms.

How Fungi Feed

Fungi are dependent on dead and living material for their growth. They obtain their nutrients in three basic ways:

- Saprobic[3] – growing on dead organic matter;
- Symbiotic – growing in association with other organisms;
- Pathogenic or Parasitic – causing harm to another organism.

The majority of wild edible fungi species are symbiotic and form mycorrhizas with trees. Saprobic edible fungi are also collected from the wild but they are best known and most widely valued in their cultivated forms. Plant pathogenic fungi cause diseases of plants and a small number of these microfungi are eaten in the form of infected host material. The different modes of feeding are shown in Plate 2 and described briefly below.

Saprobic Fungi

Fungi colonize rotting wood and organic matter found in soil. Many species cannot be seen with the naked eye (microfungi) but there are (edible) macrofungi that fruit on fallen logs and bracket fungi that grow from dead or dying parts of standing trees.

Agaricus arvensis is a commonly collected wild edible species that occurs in pastures and grassy areas. Edible species of *Favolus* are collected from dead wood inside tropical rain forests. The wild edible fungi used by the Yanomam Indians in Brazil are all saprobic and occurred in slash and burn areas where rotting wood was present.

In the wild, the volume and value of saprobic species used as food are small by comparison with the symbiotic edible fungi, though more edible saprobic species are collected. Their overall value is much higher because they are widely cultivated: a recent figure of US$18 billion was quoted for the annual, global trade in cultivated, saprobic species.

Saprobic species need a constant supply of suitable organic matter to sustain production in the wild and this can be a limiting factor in production. *Shi'itake* (*Lentinula edodes*) mushroom cultivation in one area of China is threatened by the supply of suitable tree branches from nearby forests.

Saprobic macrofungi are also highly valued for their medicinal properties. Most are cultivated, though *Ganoderma* spp. are also collected from the wild. The list of symbiotic macrofungi with medicinal properties is a short one, though there is some indication that they have been studied less because they cannot be cultivated.

Symbiotic Fungi

The most common form of symbiosis associated with wild edible fungi is that known as a mycorrhiza. Many plants depend on these fungus-root associations for healthy growth.

A special type known as an ectomycorrhiza (ECM) is found on trees growing in the Taiga in the Russian Federation and the rain forest of Borneo and includes legume trees as well as conifers. Ectomycorrhiza are typically formed by macrofungi and they include many of the key edible species that are collected in the wild, such as chanterelles (*Cantharellus* spp.) and *Amanita* species.

Table: Plant families with edible ectomycorrhizal fungi

Family	Examples
Betulaceae	*Betula* (birches)
Caesalpinioideae	*Afzelia, Brachystegia, Isoberlinia, Julbernardia*
Casuarinaceae	*Casuarina*
Cupressaceae	*Cupressus*
Dipterocarpaceae	*Shorea, Dipterocarpus, Monotes*
Euphorbiaceae	*Uapaca*
Fagaceae	*Castanea* (chestnut), *Castanopsis, Fagus* (northern beech), *Nothofagus* (southern beech), *Quercus* (oak)
"Legumes"	*Acacia*
Myrtaceae	*Eucalyptus*
Pinaceae	*Pinus* (pines), *Picea* (spruces), *Abies* (firs), *Larix* (larches)
Papilionoideae	*Pericopsis*
Nyctaginaceae	*Neea*

The mycorrhiza helps the tree to grow in nutrient-poor soils, such as the miombo woodland of central and southern Africa. A sheath of hyphae wraps around the root. They penetrate the root structure but not the actual root cells themselves, forming a living contact between the fungus and the tree. The fungus helps the tree gather water from a wider catchment and delivers nutrients from the soil that the tree cannot access. The tree provides the fungus with essential carbohydrates.

Termitomyces contains important wild edible species. These fungi only grow in association with termites and their nests and are dependent on the organic matter brought by the insects from their feeding on trees.

Although *Termitomyces* are saprobic, they are symbiotic with termites. Twenty edible species of *Termitomyces* have been recorded from Africa and Asia. They are regularly collected and also sold. *T. titanicus* is the world's largest edible fungus, although other species are much smaller.

Rural people have long associated the appearance of edible fungi with particular trees and have incorporated this in local names. In southern Africa, *chimsuku* and *kamsuku* both describe *Lactarius* spp. that grow under *masuku* trees.

Some edible ectomycorrhizal fungi produce their fruiting bodies underground. The best known examples are the truffles (*Tuber* spp.: Plate 4). Over 400 species of edible ECM have been recorded. There are also many ectomycorrhizal fungi which produce fruiting bodies that are not edible or are poisonous.

The production of fruiting bodies depends on a complex set of factors and in some years production can be negligible.

In Botswana, 14 tonnes of *Terfezia pfeilii,* one of the "desert truffles", were bought from one small community in one season; the next year only four fruiting bodies were located over a much larger area. The lack of certainty of harvests from one year to the next makes it difficult to plan commercial exploitation and some attempts have been made to overcome this by "cultivating" key mycorrhizal species such as *Tricholoma matsutake.*

Trees are successfully infected with truffles and managed under controlled conditions in Italy and elsewhere, but the time, effort and money required are only justified – assuming a good knowledge of the ecology of the fungus concerned – for the most valuable edible mycorrhizal species.

Tree species can form mycorrhizas with more than one fungus, and a fungus may associate with more than one tree. Some ECM are "native" to a region: in Madagascar an edible *Russula* grows on exotic eucalyptus. Other edible ECM have been introduced and *Boletus edulis* is now found throughout southern Africa following the establishment of pine plantations. ECM have been most intensively studied in the past on temperate tree species but there have also been steady advances on tropical ECM in Africa.

Lichens are "self-supporting" associations between fungi and an alga or cyanobacterium and are the final example of a symbiosis that has edible properties. A lichen is a biological and not a systematic group and several valuable species are eaten by people in Europe, Asia and North America and used for other economic purposes. They are not included in this book. Further information is available from a number of sources.

Plant Pathogens and Parasitic Fungi

In several countries people eat plant material infected with plant pathogenic fungi. Maize cobs infected with the smut fungus *Ustilago maydis* are consumed in large quantities in Mexico, both fresh and canned. They are known locally as *huitlacoche* or *cuitlacoche. U. maydis* is a microfungus: it does not form a visible fruiting body and the only signs of its presence are a mass of dark spores. The cobs appear to become sweeter as the result of fungus attack, and similar changes have been noted for the edible rust fungus *Cronartium conigenum* on pines in Mexico.

Other examples include: *Ustilago esculenta* on wild rice; *Sporisorium cruenta* on sorghum in China; winged bean infected by *Synchytrium psophocarpi* in Indonesia.

Hypomyces lactifluorum is a parasite macrofungus that grows on other macrofungi (boletes). It is eaten from Canada through to Guatemala and completes the range of ecological niches occupied by wild edible fungi.

NON-MYCOTOXIC FUNGI

Fungi are a major cause of spoilage in stored grain. The Food and Agriculture Association estimates that 25% of the worlds food crops are affected by mycotoxins (the by-products of fungal growth) during growth and storage. The damage of fungi is second only to that caused by insects in stored grain products. Many of the fungi cause damage to the crops themselves with little or no toxic effects on humans and animals.

Common Smut or Bunt. Common smut is caused by two fungi; *Tilletia tritici and Tille-tia laevis.* Because it requires cool moist soil conditions, the disease is less of a problem on spring planted wheat than winter wheat. Common smut reduces wheat yields and grain quality. Wheat contaminated with bunt spores has a pungent, fishy odour and a darkened appearance. Wheat that has an unmistakable odour of smut will be designated "light smutty" on official inspection certificates. In addition, bunt spores released during combining are combustible, and have caused explosions and fires during harvesting with mechanical harvesters.

Plants may be moderately stunted but are not easily distinguished until the heads emerge. Bunted heads are slender and maintain their green colour longer than healthy heads. The glumes may spread apart exposing the smut balls they contain. Smut balls are approximately the shape of normal kernels and are dull gray-brown. The small balls often rupture at harvest, releasing black, powdery spores. Wheat containing smut balls will be designated as "light smutty" (6-30 smut balls) or "smutty" (31 or more smut balls) on official inspection certificates. Dwarf Bunt or TCK Smut. The *Tilletia* species of fungi cause smut or bunt diseases in wheat, rye, and barley. *Tilletia controversa,* commonly known as Dwarf Bunt or TCK Smut, causes dwarfing or stunting in the growth of the plant itself along with reduced yields. The major source of transmittal is by the spores laying dormant in the soil until the next crop year. These spores can lay dormant for up to ten years retaining their infectious properties.

Karnal Bunt. Karnal Bunt is thought to have originated in India (therefore the name *Tilletia indica*) but has since been discovered in many places worldwide. It is thought to be in Afghanistan, Lebanon, Mexico and now the United States. Karnal Bunt has spores that are primarily wind-borne, but infection can occur through planting in contaminated soil, use of infected seed, and coming into contact with contaminated equipment and machinery.

The spores attack the developing kernels within the seed head with little or no outward sign of infection. The infected kernels are shrunken at the germ end and are covered with sori (small hairlike filaments) that will discolour flour made from infected kernels. These kernels will also impart a fishy odour and taste to flour making their commercial use impractical.

Black Tip / Black Point. Black Tip fungus is another non-mycotoxic fungus that attacks wheat and barley. The name Black Tip or Black Point can generally refer to any of a number molds that form dark brown to black sooty mold. The principal species of infection are *Alternia, Fusarium, and Helminthosporium,* with the species *Helminthosporium* generally being the causal agent in infections associated with wheat.

The mold generally occurs when seeds are sown in infected soils, but can also occur when the seeds themselves are infected. The most common forms of *Helminthosporium* are generally associated with dry, warm soils. Moisture in the form of rain or extended periods of high humidity (over 90% relative humidity) then start the disease on the maturing kernels. If the mold begins early in kernel growth sterility and germ death occur, often accompanied with a shrivelled germ resulting in decreased yields. If the mold attacks in the later stages of kernel development the seed is discoloured and often carries an odour causing market discounts. These molds are also associated with seedling blight and root rot which will effect the overall health of the plant leading to increased susceptibility of other plant diseases and molds.

Blue-Eye Rot. Blue-eye mold occurs in stored corn with high moisture content. Blueeye damage is caused by species of *Penicillium* and is characterized by a blue-green discoloration in the germ area.

The discoloration results when *Penicillium* fungi invade the germ area through the tip of the kernel. Some corn varieties have a purple coloured plumule which can be mistaken for the presence of *Penicillium* fungi in the germ area. Corn Smut. Smut is caused by *Ustilage maydis* and is always present in field corn. No harmful effects have been noted from feeding silage made from smutty corn to livestock.

Background. The principal damage caused by the fungi was the mold. There were little or no toxic affects from by-products of the mold. The molds were also limited as to the hosts they attacked. Mycotoxins on the other hand are metabolites (by-products) of the growth of the molds. They have very real toxic side effects to other plants, animals, and humans. They are also generally less selective of the hosts they attack and can cross plant species.

The species *Fusarium* can and will attack both corn and wheat with different effects in each plant. In wheat, they cause scab damage to the kernels and produce deoxynivalenol (DON). In corn, they create Gibberella Ear Rot and produce DON, zearalenone (ZEN) and T-2 toxins. Both DON and ZEN have toxic effects on animals and humans, with differences depending on the species. Mycotoxin contamination of crops has been a world wide problem for thousands of years. Only in the last thirty or forty years has technology allowed researchers to isolate the fungal mycotoxins and study the effects on feed crops, livestock, and their affects on humans.

The study of mycotoxins is a narrow part of the research into naturally occurring toxins and their effects on plants, animals, and humans. Much of the information in print becomes quickly dated as the field expands and grows. With recent improvements in testing methods more research is now being done, and in greater detail, with lower costs than was previously possible.

Many of the newest research papers are now posted on the Internet allowing new information to reach more people faster and more efficiently than was possible before. Information that was published only 20 to 25 years ago is now considered dated and questionable. But with the limited nature of the end usefulness of the information all information must be considered.

The problems associated with mycotoxin contamination of grains are world wide and are uncontained by national borders. A look at a map of North America shows that the important grain producing areas stretch from north central Canada to the southern reaches of the United States. These areas in turn export their products to countries in Asia, the Pacific islands, South and Central America, Europe, the Middle East, India and Africa.

Scientists estimate that there are 300 to 400 mycotoxins presently identified with more being isolated as new techniques and processes evolve. The most frequently found mycotoxins are aflatoxin, deoxynivalenol (DON), zearalenone (ZEN), fumonisin, and T-2 as far as grain crops are concerned. In surveys conducted by the North Carolina Cooperative Extension service in 1990 some amount of aflatoxin, DON, or fumonisin was found in over 70% of the samples tested.

Mycotoxic molds generally attack the kernels of grain robbing the nutrients and lower the fat, protein, and vitamin content of the grain.

WHAT ARE MYCOTOXINS?

The mold also often changes the colour of the kernels, the consistency (texture), and often imparts an odour that causes feed refusal in animals. These effects lead to economic losses due to impaired health in animals and humans, reduced productivity (reduced production of eggs, milk, and weight gain), and in severe cases death to animals and humans. When the grain is processed into final products like flour or feed, the visible mold may be removed, but the majority of toxins are not and can still cause poisoning.

Economic Effects

The economic effects attributed to mycotoxin infection are widely felt in all sectors of the production and consumption of grain products. Grain producers are affected by limited yields, restricted end markets, and price discounts. Grain handlers are affected by restricted storage options, cost of testing grain lots, and loss of end markets. Grain processors incur higher costs to due to higher product losses, monitoring costs, and restricted end markets.

Consumers end up paying higher end product prices due to increased monitoring at all levels of handling, and in extreme cases health problems due to consumption of contaminated products. Societies as a whole end up paying higher costs due to increased regulations, needed research, lower export costs, and higher import costs. While these costs are found at every level of the grain production system, it is almost impossible to put a dollar figure on the losses.

Estimates of losses to small portions of the grain and related industries are the best that can be accomplished. The North Carolina Cooperative Extension Service published an estimate for 1992 that the losses to the animal production industry in North Carolina were $20 million for poultry, $10 million for swine, $5 million for dairy, $1 million for beef and sheep, and $1 million for horses. In 1990 a vomitoxin (DON) outbreak in New York and other Northeastern states, the Northcentral U.S., and Eastern Canada had widespread economic affect. The New York Corn Growers estimated, conservatively, that $12 million dollars was lost by the corn farmers due to lost markets, decreased crop value, and costs associated with testing for the 1990-91 crop year.

Health Hazards. The health hazards associated with mycotoxin contamination in humans are rarely seen in North America and Europe. This is generally attributed to a higher level of general health than is seen in underdeveloped countries and better control of food and feed storage.

The levels of intake of affected products that are necessary to bring about poisoning in healthy individuals are actually quite high. In the rare cases that humans have been reported to be poisoned by mycotoxins, the populations were consuming limited quantities of other non-tainted foods and lived in areas of economic and environmental stress. The greatest threat of health hazards to humans comes from long term exposures of tainted food products, either from spoilage or from consuming milk or meat from animals that have been fed contaminated feed. One possible avenue of concern to humans is the suspected link between aflatoxin and cancer. While the International Agency for Research on Cancer (IARC) has listed aflatoxin B1 as having a definite link to cancer in animals, it is listed as having a *probable* link to cancer in humans. The studies that have been done on the cancer link to humans have been done in Africa and Asia and show an association between aflatoxin and cancer but no definitive cause and effect relationship has yet been documented.

Mycotoxicoses (poisoning due to mycotoxins) have several common symptoms that are shared from species to species and toxin to toxin. These symptoms include:

1. Drugs and antibiotics are not effective in treatment.
2. The symptoms can be traced (associated) to food or feedstuffs.
3. Testing of food/feedstuffs reveals fungal activity.
4. The symptoms are not transmissible to control subjects.
5. The degree of toxicity in subjects is influenced by age, sex, and the nutritional status of the host.
6. Outbreak of symptoms is seasonal. Mycotoxins have been linked to birth defects in many animals, nervous system problems (tremors, limb weakness, staggering, and seizures), and tumours of the liver, kidneys, urinary tract, digestive tract, and the lungs.

Aflatoxin primarily attacks the liver, with secondary effects shown in decreased production, and immune system suppression. The trichothecene group of mycotoxins (DON and T-2) cause necrosis and hemorrhage of the digestive tract with decreased blood production in the bone marrow and spleen along with changes to the reproductive system as the secondary sites of attack.

Zearalenone has affects different from other mycotoxins in that it mimics the bodies production of estrogen causing feminization of male animals and interference with conception, ovulation, and fetal development in female animals.

One of the primary similarities in mycotoxicosis is the health status of the host before infection, which will to a large extent determine the degree to which the host is attacked. In both animals and humans if the host is healthy the final prognosis is generally very good except in cases of acute poisoning where very large doses of toxin are eaten over a very short period of time.

Many variables affect the degree of susceptibility of various hosts. These include:

General health. A healthy individual is more able to fight the toxins than a one that starts out malnourished or diseased.

Age. The very young and very old have weakened immune systems that are less able to fight the effects of toxicoses.

Sex. In general female animals, and to a degree female humans, seem to be more susceptible to the effects of mycotoxicoses.

Environment. Hosts exposed to harsh living conditions of neglect or squalor have an added burden on their systems.

Adequate food storage. If grain is left in open storage to the effects of weather the grain is further weakened allowing for continued fungal growth. If the grain is also not adequately dried, the conditions for fungal growth continue to persist.

Exposure Level. Very high doses of aflatoxin attack the host quicker than lower doses.

Exposure duration. A very short time of exposure (a single dose) allows the host time to fight the infection where a continued exposure tends to have cumulative effects.

Other food sources. If contaminated grain is the only or main source of nourishment, the host has less nutrients available to feed their system.

Lack of Regulatory and Monitoring systems

Areas that lack the means to regulate and monitor the presence of mycotoxins in general, and aflatoxin in particular, leave their inhabitants open to unknown poisonings that can continue over long periods of time unchecked. Although the mycotoxins have been greatly researched over the course of the last forty years, little research has been done on the interactions of the mycotoxins and their combined effects. Almost all research has been in determining the effects of pure strains developed in the laboratory.

Animal feed can contain several different grains and grain in storage can contain several different strains of the various mycotoxins. Studies done with natural field contaminated DON and ZEN have produced results that have varied from those that have been carried out in the laboratory. This leads researchers to believe that there are unknown strains of toxins in the field or that the toxins interact with each other to produce effects greater than or different from what laboratory tests have predicted would occur.

Regulatory Control. In 1965, the Food and Drug Administration (FDA) established action limits of 20 parts per billion (ppb) of aflatoxin in all food and feeds to limit the inclusion of this contaminate into the food chain.

But since aflatoxin can occur both in raw products and in finished by-products, this has necessitated the testing for aflatoxin and other mycotoxins at many points in the food chain. As technology has progressed in recent years testing has become simplified, faster, and cheaper (using ELISA methods) allowing more testing to be accomplished in all phases of food handling.

The Grain Inspection; Packers and Stockyards Administration (GIPSA) tests all corn that is exported for aflatoxin with an action level of 20 ppb. If a sample tests above 20 ppb, the FDA must be notified. Testing for vomitoxin is done as a service to customers with no FDA action level set. FGIS will institute new testing services as the tests are certified for reliability and repeatability, and as customer demand requires.

Aspergillus Toxins

Aspergillus is an important genus in foods with most species occurring as spoilage or biodeterioration fungi. *Aspergillus* is a large genus containing more than 100 recognized species, several of which are capable of producing mycotoxins.

Nearly 50 species of *Aspergillus* have been listed as producing toxic metabolites. Those of greatest significance in feed and foods include: aflatoxin, ochratoxin A, sterigmatocystin, cyclopiazonic acid, citrinin, patulin, and tremorgenic toxins. *Aspergillus* species produce toxins that exhibit a wide range of toxicities, with the most significant effects being long term. Aflatoxin B1 is a potent liver carcinogen. Ochratoxin A and citrinin both affect kidney function. Cytopiazonic acid has a wide range of effects and tremorgenic toxins affect the nervous system.

While there is a known link between aflatoxin and cancer in animals, it should also be noted that the *Aspergillus* species of fungi also have many beneficial uses. One of the largest commercial uses of *Aspergillus* fungi is in the production of soft drinks. The extraction of pure citric acid from fruits and vegetables has proved to be to expensive, so the manufacturers have developed a way to use large vats to ferment *Aspergillus niger* to form artificial citric acid.

The Japanese have developed methods of fermenting rice with strains of *Aspergillus flavus* (aflatoxin) to cause the enzymes in the kernels to breakdown the carbohydrates into simple sugars, to produce a sweetened rice drink. These drinks or Koji, are marketed under many different brand names.

The family of products that includes miso, soy sauce, and sake also use strains of *Aspergillus oryzae* to ferment grain products into usable food and drink products.

Penicillium Toxins

Penicillium is a large genus with over 150 species recognized and at least 50 species of common occurance. The discovery of penicillin in 1929 gave impetus to a search for other *Penicillium* metabolites with antibiotic properties. This search led to the recognition of "toxic antibiotics" or mycotoxins. Nearly 100 *Penicillium* species have been reported as toxin producers. Of these the following nine mycotoxins produced by 17 *Penicillium* species are potentially significant to human health: citreoviridin, citrinin, cyclopiazonic acid, ochratoxin A, patulin, penitrem A, PR toxin, Roquefortine C, and Secalonic acid D.

The toxins produced by *Penicillium* species can be placed in two general groups: those that affect the liver and kidney function, and those that are neurotoxic. The *Penicillium* toxins that affect liver or kidney function are asymptomatic or cause generalized debility in humans or animals while the neurotoxins are characterized by sustained trembling.

Fusarium Toxins

Fusarium species are the most important group of mycotoxigenic molds other than *Aspergillus* and *Penicillium*. Many *Fusarium* species are plant pathogens and most can be found in the soil.

Fusarium species are most often encountered as contaminants of cereal grains, oilseeds, and beans. Corn, wheat, barley and products made from these grains are most commonly contaminated although rye, triticale, millet, and oats can also be contaminated.

F. graminearum is a plant pathogen found worldwide in the soil and is the most widely distributed toxigenic *Fusarium* species. It causes various diseases of cereal grains such as gibberella ear rot in corn and head blight or scab in wheat and barley. The mycotoxins produced by *F. graminearum* include:

deoxynivalenol, zearalenone, 3-acetyldeoxy-nivalenol, 15-acetyldeoxynivalenol, diacetyldeoxynivalenol, nivalenol, T-2, neosolaniol, and diacetoxyscirpenol. *F. moniliforme* is a soilborne plant pathogen that is found in corn growing in all regions of the world. It is the most prevalent mold associated with corn. It has also been found in rice, sorghum, yams, hazelnuts, pecans, and cheeses.

F. moniliforme has long been suspected of being involved in animal and human diseases. Animal diseases associated with *F. moniliforme* include equine leukoencephalomalacia (ELEM) a liquefactive necrosis of the brain of horses, pulmonary edema and hydrothorax in swine, liver cancer in rats, and abnormal bone development in chicks and pigs.

The main human disease associated with *F. moniliforme* is esophageal cancer. Several studies have linked the presence of *F. moniliforme* and fumonisins in corn to high incidences of esophageal cancer in humans in certain regions of the world including an area around Charleston S.C. Mycotoxins that have been associated with *F. moniliforme* include fumonisins, fusaric acid, fusarins, and fusariocins.

Alternaria Toxins

Alternaria species may be significant as potential contaminants of food. *Alternaria* infects the plant in the field and may contaminate wheat, sorghum, and barley. *Alternaria* species also infect various fruits and vegetables and can cause spoilage of these foods in refrigerated storage.

Alternaria toxins include: alternariol, alternariol monomethyl ether, altenuene, tenuazonic acid, and the alertoxins. Little is known about the toxicity of these toxins; however, cultures of *Alternaria* that have been grown on corn or rice and fed to rats, chicks, turkey poults, and ducklings have been shown to be quite toxic.

Claviceps Toxins

The ergot mold, *Claviceps purpurea,* is the cause of the earliest recognized human mycotoxicosis, ergotism. Ergot has been reported in sporadic outbreaks in Europe since 857, with near epidemic outbreaks in the Middle Ages. Ergot is a disease of cereal grains such as rye and wheat in which the grains are replaced by ergot sclerotia that contain toxic alkaloids. The main ergot alkaloid, ergotamine, has vasoconstrictive properties that can cause swollen limbs, and alternating burning and cold sensations in the fingers, hands, and feet (St. Anthony's Fire).

Aflatoxin

Pathogen. The name aflatoxin comes from A(Aspergillus) + FLA(flavus) + toxin. Modern research into aflatoxin had its beginnings in 1961, looking into what caused the deaths of 100,000 young turkey poults in England. The research traced the poison to contaminated Brazilian peanut meal that had been used as feed. When the feed was given to ducks and pheasants, the same outcome was produced. Research found that there were four different metabolites formed from aflatoxin. When the contaminated grain in question was viewed under a black light, the metabolites glowed either blue (B metabolite) or green-yellow (G metabolites). Of the two distinct colour varieties there were isolated two distinct toxins (B1, B2, G1, G2). The subscripts refer to their separation patterns on TLC plates. Of the four metabolites B1 was the most predominate and the most toxic.

Further research in the years since has found other metabolites, with two found in the milk (M1 and M2) and urine of lactating mammals. Aflatoxins M1 and M2 are produced from their respective B aflatoxins by hydroxylation in lactating animals and are excreted in milk at the rate of approximately 1.5% of the rate of ingested B aflatoxins. Research has also found that aflatoxin is most commonly found in tree nuts, peanuts, and oilseeds including corn and cottonseed.

Ecology. The greatest problems associated with aflatoxin are in corn production and foodstuffs. These problems occur for two reasons. Corn is grown in climatic areas that give the fungi/mold the greatest opportunity for growth and dispersal, and the areas that grow corn consume it as a main part of the diets of both animals and humans.

Aflatoxin grows best at temperatures of 80E to 90E F, but can survive at temperatures as low 40E F. The mold also needs a high moisture content in the host, either through kernel moisture or in the form of rainfall. Kernel moisture of 20% or greater is optimal, but the fungus can survive and grow in grain with moisture content as low as 15%. In corn, insect damage to developing kernels allows entry of aflatoxigenic molds, but invasion can also occur through the silks of developing ears. Research conducted into the growth of aflatoxin has indicated that the fungus needs some form of associated stress in the plants for the fungus to invade. The stress may be in the form of drought that weakens the plant system, extended periods of high temperature, damage from insects or birds, high crop density, or competition from weeds. All of these conditions weaken the host or provide a means of entry to the spores to establish a foothold in/on the host.

CAUSES OF PLANT DISEASE

Plant diseases can be infectious (transmitted from plant to plant) or noninfectious. Noninfectious diseases are usually referred to as disorders. Common plant disorders are caused by deficiencies in plant nutrients, by waterlogged or polluted soil, and by polluted air. Too little (or too much) water or improper nutrition can cause plants to grow poorly. Plants can also be stressed by weather that is too hot or too cold, by too little or too much light, and by heavy winds. Pollution from automobiles and industry, and the excessive application of herbicides (for weed control) can also cause noninfectious plant disorders. Infectious plant diseases are caused by pathogens, living microorganisms that infect a plant and deprive it of nutrients. Bacteria, fungi, nematodes, mycoplasmas, viruses and viroids are the living agents that cause plant diseases. Nematodes are the largest of these agents, while viruses and viroids are the smallest. None of these pathogens are visible to the naked eye, but the diseases they cause can be detected by the symptoms of wilting, yellowing, stunting, and abnormal growth patterns.

Bacteria

Some plant diseases are caused by rod-shaped bacteria. The bacteria enter the plant through natural openings, like the stomata of the leaves, or through wounds in the plant tissue. Once inside, the bacteria plug up the plant's vascular system (the vessels that carry water and nutrients) and cause the plant to wilt. Other common symptoms of bacterial disease include rotting and swollen plant tissues. Bacteria can be spread by water, insects, infected soil, or contaminated tools. Bacterial wilt attacks many vegetables including corn and tomatoes, and flowers. Crown gall, another bacterial plant disease, weakens and stunts plants in the rose family and other flowers. Fireblight attacks apple, pear, and many other ornamental and shade trees.

Fungi

About 80% of plant diseases can be traced to fungi, which have a great capacity to reproduce themselves both sexually and asexually. Fungi can grow on living or dead plant tissue and can survive in a dormant stage until conditions become favourable for their proliferation. They can penetrate plant tissue or grow on the plant's surface. Fungal spores, which act like seeds, are spread by wind, water, soil, and animals to other plants.

Warm, humid conditions promote fungal growth. While many fungi play useful roles in plant growth, especially by forming mycorrhizal associations with the plant's roots, others cause such common plant diseases as anthracnose, late blight, apple scab, club root, black spot, damping off, and powdery mildew. Many fungi can attack are variety of plants, but some are specific to particular plants.

The list of fungi and the plants they infect is a long one. Black spot attacks roses, while brown rot damages stone fruits. Damping off is harmful to seeds and young plants. Downy mildew attacks flowers, some fruits, and most vegetables. Gray mold begins on plant debris and then moves on to attack flowers, fruits, and vegetables. Oak root fungus and oak wilt are particularly damaging to oaks and fruit trees. Peach leaf curl targets peaches and nectarines. Powdery mildew, rust, sooty mold, and southern blight attack a wide variety of plants, including grasses. Texas root rot and water mold root rot can also infect many different plants. Verticillium wilt targets tomatoes, potatoes, and strawberries.

Viruses and Viroids

The viruses and viroids that attack plants are the hardest pathogens to control. Destroying the infected plants is usually the best control method, since chemicals to inactivate plant viruses and viroids have not proven effective. While more than 300 plant viruses have been identified, new strains continually appear because these organisms are capable of mutating. The symptoms of viral infection include yellowing, stunted growth in some part of the plant, and plant malformations like leaf rolls and uncharacteristically narrow leaf growth. The mosaic viruses can infect many plants. Plants infected with this virus have mottled or streaked leaves; infected fruit trees produce poor fruit and a small yield.

Nematodes

Nematodes are tiny microscopic animals with wormlike bodies and long, needlelike structures called stylets that suck nutrients from plant cells. They lay eggs that hatch as larvae and go through four stages before becoming adults. Nematodes have a 30-day life cycle, but they can remain in a dormant state for more than 30 years. Nematicides are chemicals used to control nematode infestations. Marigolds are resistant to nematodes and are often planted to help eliminate them from infected soil.

Nematodes primarily attack plant roots, but they may also destroy other parts of the plant either internally or externally. They thrive in warm, sandy, moist soil and attack a variety of plants including corn, lettuce, potatoes, tomatoes, alfalfa, rye, and onions. However, all nematodes are not harmful to plants. Some are actually used to control other plant pests such as cutworms, armyworms, and beetle grubs.

DEFINITION OF PLANT PATHOLOGY

The study of plant diseases is known as plant pathology. Infectious diseases are caused by living organisms called pathogens. Noninfectious diseases caused by environmental stress and damage by weather and other environmental factors also will be covered. Indirectly, environmental factors that cause a plant to be stressed may result in the plant's gradual decline. Decline results in the plant being more susceptible to disease organisms. Because of this, diagnosing plant diseases can be tricky. The real cause of a problem may be the stress factors, with the disease simply being a secondary factor.

DISEASE TRIANGLE

Three critical factors or conditions must exist for disease to occur: a Susceptible Host Plant, a Pathogen, and the right mix of environmental conditions. The relationship of these factors is called the disease triangle. If only a part of the triangle exists, disease will not occur. Understanding the disease triangle helps us understand why most plants are not affected by the many thousands of diseases that exist.

Pathogen

Pathogens are microorganisms that cause disease. Because they are living, they are called Biotic (bye-AH-tick) agents or causes. Pathogens can be Fungi (Fun-geye), Bacteria, Viruses, Mycoplasmas (MY-crow-plas-mahs) or Nematodes (Knee-ma-toads). Each has a different life cycle, which includes an infectious stage. Most pathogens are host-specific to a particular plant species, genus or family. For instance, blackspot of rose will not attack marigolds or lettuce. Some diseases, such as the powdery mildews, produce similar SYMPTOMS on different plants. However, the fungi involved are usually host-specific. The rose powdery mildew fungus will not infect zinnias or turfgrass or vice-versa.

Susceptible Host

A susceptible host has a genetic makeup that permits the development of a particular disease. The genetic defence against a disease is called disease resistance. This resistance can be physical characteristics of the plant (fuzzy or waxy leaf surfaces), chemical characteristics (enzymes that kill pathogens and lack of enzymes) and growth patterns (ability to block off diseased tissue or outgrow damage).

Plants also may be disease-tolerant. Even though infected with a disease, they can grow and produce a good crop or maintain an acceptable appearance. The plant outgrows the disease and symptoms are not apparent or at a damaging level. It is important to remember that plants labelled as disease-resistant are resistant only to a particular disease. They are not resistant to all diseases. Resistance does not mean immunity. Under extreme circumstances, resistant plants may be infected by the disease to which they have resistance. For disease to occur, the host plant must be at a stage of development that allows it to be susceptible to infection. For example, damping-off only affects seedlings. Botrytis is primarily a disease of buds, although it also can occur on flowers and leaves. Also, it is important that the pathogen be in a proper stage of its development to infect host plants.

Environmental Conditions

Certain environmental conditions must exist for disease pathogens to cause infection. The specific conditions vary for different pathogens. High moisture and specific temperature ranges, for example, are necessary for many fungal diseases. These conditions must continue for a critical period of time while the pathogen is in contact with the host for infection to occur. Moisture, temperature, wind, sunlight, nutrition and soil quality affect plant growth. If one of these factors is out of balance for the culture of a specific plant, that plant may have a greater tendency to become diseased. For example, lilacs growing in shade are more likely to be infected with powdery mildew than those growing in full sunlight. Environmental conditions also affect the growth and spread of disease pathogens. Very dry or wet weather will have an accompanying set of diseases that thrive under these conditions.

Moisture

Moisture in the plant environment can include humidity, dew, rainfall or water from irrigation. Moisture is critical to the spread of most plant diseases. Familiar diseases, such as black spot, fireblight and apple scab require moisture to spread to and infect new host plants.

Constantly wet foliage from overhead watering is a condition that promotes disease development. Seedlings grown indoors in soggy, unsterilized potting medium and pots are more prone to damping-off, a fungal disease.

Temperature

Each disease pathogen has a specific temperature range for growth and activity. There are warm-weather and cool-weather diseases. Many powdery mildew diseases are late summer, warmer temperature diseases. Temperature affects how rapidly pathogens multiply.

Soil temperature can also be critical for disease infection. Cool, wet soils promote fungal root diseases. Temperature extremes can cause stress in host plants, increasing susceptibility.

Wind and Sun

The combination of wind and sun affects how quickly plant surfaces dry. Faster drying generally reduces the opportunity for infection. Wind can spread pathogens from one area to another, even many miles. Wind and rain together can be a deadly combination. Windblown rain can spread spores from infected plant tissue, blowing these pathogens to new host plants.

Sunlight is very important to plant health. Plants that do not receive the right amount of sunlight to meet their cultural requirements become stressed. This may make them more susceptible to infection.

Soil and Fertility

Soil type can affect plant growth and also development of some pathogens. Light sandy soil low in organic matter favours growth of many types of nematodes. Damping-off disease increases in heavy, cold, waterlogged soils. Soil pH affects pathogen development in some diseases. Clubroot of cabbage occurs in soils with a low pH, for example. High soil pH is a factor in the development of scab on potatoes. Fertility affects a plant's growth rate and ability to defend against disease. Excessive nitrogen fertilization can increase susceptibility to pathogen attack. It causes formation of SUCCULENT tissue and delays maturity. This can contribute to certain patch diseases in lawns. Nitrogen deficiency results in limited growth and plant stress which may cause greater disease susceptibility.

PLANT PATHOGENS

To help you develop a "disease vocabulary," here are some of the common pathogen groups. Pathogens are grouped taxonomically into related groups.

It is helpful to know the type of pathogen in order to prescribe a control. Specific diseases are discussed in the chapters dealing with the plant. Plant disease names are often very colourful and descriptive, as in fireblight, black knot and clubroot. Many disease names indicate something about the appearance of the symptoms.

Fungi

Fungi (plural of fungus) are the largest group of plant pathogens. Fungi often develop into colonies, like bread mold. Some fungi develop into large structures, such as mushrooms. Fungi spread by spores carried by wind, water or animals. They enter host plants directly or through natural openings or wounds. They cause damage by producing substances that change or destroy plant tissues.

Common fungal diseases include powdery mildew, rust, leaf spot, blight, root and crown rots, damping-off, smut, anthracnose, and vascular wilts.

Powdery Mildews

These fungal diseases are very HOST-SPECIFIC. It is common to see powdery mildew on lilacs, roses, zinnias, melons, beans, cucumbers, turfgrass and many other plants. The symptom is a white or gray, powdery growth on leaves and stems. Powdery mildew will not usually kill a plant; however, it may weaken plants, and it is unsightly. Providing adequate sunlight and air circulation to lower relative humidity and using fungicide sprays for high-profile plants help control powdery mildew.

Rust

Rusts are also host-specific. They produce masses of easily noticed, orange or dark red spore masses on leaf tissue. Rusts are commonly seen in our region on turfgrass, hollyhocks, snapdragons, hawthorns and the small fruits. Rusts usually develop during cool weather. They are spread by wind and splashing water. Rusts require water to reproduce and infect host plants. Controls include watering early in the day and applying protective fungicidal sprays.

Leaf Spot and Blight

Rose growers are very familiar with black spot, a common leaf-spotting disease of roses. Scab on crabapple and pyracantha are leaf-spotting diseases that can cause defoliation. Botrytis flower blight occurs on annuals and some perennials during wet seasons. Anthracnose of shade trees is a blight frequently seen in Ohio on sycamores, maples, ashes and oaks. Fungi-causing tree diseases are spread by wind and splashing water. Grow resistant cultivars, maintain plant vigour, enhance leaf drying by proper pruning and water early in the day to help avoid leaf spotting and blights. Fungicides provide protection from infection when applied before infection by fungal spores.

Root and Crown Rots

These diseases, which include water mold diseases, attack a wide variety of plants. They cause damage to roots and crowns of plants. Damage to some of the root system results in poor growth, yellowing or stunting of the plant. Root rot fungal pathogens are found in almost all soils. However, they do not survive as well in well-drained soils. These organisms can live in soil for years in a dormant state. Proper cultural practices, such as correct planting depth and improving drainage, are important controls.

Stem and Twig Cankers

Cankers may be compared to sores on a stem or twig. The tissue may be raised or sunken, discoloured or split open. Twig cankers on junipers and leaf and stem blight of pachysandra are common examples. Most of the canker diseases are not controllable with fungicides. Maintaining plant vigour, protecting from injury, removing diseased parts and growing plants suitable for the site help prevent canker problems.

Vascular Wilts

Dutch elm disease is the classic wilt disease. The fungus invades the vascular system of the plant and clogs it, stopping the flow of water and nutrients. Verticillium wilt is common on trees, shrubs, vegetables and fruits. Discoloration of the vascular tissue sometimes can be seen when you cut diagonally through a branch. There are no effective chemical controls for wilts. Growing resistant species, maintaining plant vigour and sanitation provide the only controls.

MYCOTOXICOLOGY

Mycotoxicology is the branch of mycology that focuses on analysing and studying the toxins produced by fungus, known as Mycotoxins. As many microorganisms, fungus produce toxins as a response of biological stress in the environment, like lower nutrients or competitions for those available, conditions that trigger the secondary metabolism.

Under this secondary path the fungus produces a wide array of compounds, usually enzymes, produced them in order to gain some level of advantage, like incrementing the efficiency of metabolic process to gain more energy from less food, or attacking another microorganism and being able to use their remains as a food source.

Mycotoxicology is important in industrial processes that involves the production of food (for human or animal consumption) via the use of fungus and yeast, or with machinery in contact of these microorganisms because many of these enzymes manage to survive sterilization and can be harmful for human and animals.

DEVELOPMENT OF MYCOLOGY

Modern plant pathology did not develop alone; it grew along with and was influenced by related sciences. One of the most important of the allied fields was mycology. The fleshy fungi were,thought by the Greeks and Romans to arise de novo from dead organic matter under the influence of the elements. This idea" came naturally since all the early philosophers believed firmly in the idea of spontaneous generation of lower animals and plants. We shall find that the tenacity with which this theory was I held until the middle of the nineteenth century was a dominant factor in the delay of the development of our present concepts of parasitism and disease.

In the sixteenth, seventeenth, and eighteenth centuries some prominent it botanists were skeptical of spontaneous generation. Among these were Malpighi, the pioneer in plant anatomy, and Jung, pioneer plant systematists. On the other side were Cesalpino, another pioneer plant systematist, and Hooke, an early microscopist. The discovery of bacteria by the Dutch draper, Leeuwenhoek, seems not to have influenced current thought on reproduction of lower organisms.

A contribution which stands out in this period, is the work of Micheli. This Italian botanist studied both the higher and lower forms of fungi. In his "Nova plantarum genera" he described many new genera and illustrated their reproductive structures. He was the fist to collect spores of fungi, sow them on organic substrate and watch the development of mycelium together with sporangia and spores characteristic of the species. He was convinced that fungi arose from their own spores. However, even these careful observations and the conclusions inductively arrived at were not accepted generally. In 1748, Needham, an English clergyman and naturalist, upheld strongly the theory of spontaneous generation. To prove his point, he boiled meat in flasks, corked them, and set the containers away for a time, only to find that the contents had spoiled due to the microorganisms. In 1775, Spallanzani, an Italian investigator, challenged Needham and set out to prove that decay meat by bacteria and other lower forms of life could be prevented by heating tnematerial in a flask which was sealed off in such a way as to exclude contamination from the air. His experiments supported this contention, but others failed to secure similar results consistently. These failures, we can now understand, were due to what would presently be regarded as faulty technique. This pioneer investigator likewise did not receive recognition and was successfully refuted for the time by his opponents.

The predominant attention of botanists from the time of Cesalpino was directed towards naming and describing plants rather than searching into the phenomena whereby- they grew and developed. By the turn of the eighteenth century the dogma of the constancy of species had become deeply ingrained in the minds of men, and this concept was fostered by the ecclesiasts, who found it to cosicide well with the prevalent religious teachings. For 150 years or more, men had struggled to find a rational basis of naming and classifying plants. Linnaeus was the leader in the eighteenth century who established the presently used Latin binomial system, his work culminating in the publication of "Species plantarum" in 1753. This Swedish botanist had a wide influence. His students and followers were collectors and classifiers, but they were not experimenters. Micheli had been interested in morphology and taxonomy of the fungi; but he had also given attention to their physiology and reproduction. He was ahead of his time. Linnaeus was a product of scholasticism and a firm believer in the dogma of constancy of species. He assented to the view that fungi arose through spontaneous generation.

Persoon worked chiefly in France; Schweinitz in eastern United States; Fries in Sweden; Nees 4:9 von Esenbeck in Germany; Leveille in France. Persoon's "Synopsis methodica fungorum", pub1ccu: and lished in 1801, and Fries's "System a mycologicum", published in the opr. years 1821 to 1832, became the foundation stones of the modern application of the Linnaean system to the taxonomy of the fungi. In the main, the these men were concerned with the naming of fungi and with their class and they were not concerned with their reproduction and with the their relation to plant disease. Although Persoon gave the name Puci ke Ecinia graminis to the fungus associated with the black stem rust of wheat and, in 1794, he did not regard the fungus as the incitant of the disease. In 1818 he expressed the opinion that some fungi grew from spores, but that SU others arose by spontaneous generation.

In 1807, Prevost observed germination of the spores of the wheat bunt organism and conceived the idea that this organism penetrate the young wheat plant and was the actual cause of the disease. His revolutionary idea was refuted and did not gain favourable attention until 40 ad years later.

Although predominant attention was given by most my cologists to taxonomy, there began about 1820 a renewed interest in the type of study that Micheli had undertaken 100 years earlier.

Ehrenberg brought together known information on propagation of fungi and attributed further observations of his own on spore germination, as well as al the first description of sexuality in fungi. Nees von Esenbeck repeared al Micheli's experiment by sowing bread-mold spores on bread and later obtaining ripe sporangia. Dutrochet connected the sporophores of fleshy fungi with the subterranean mycelial strands which had been a considered up to that time as distinct fungi.

Premodern History

During the seventeenth century, plant diseases were increasing in importance in Europe with increase in-population and in intensity of plant culture. Farmers, by and large, accepted them as inevitable concomitants of unfavourable soil and climate. Some growers, however, tried to do something to alleviate the damage, for we know that treatment of wheat seed with salt brine to control bunt began in that century. The association of the barberry with stem rust of grain had, been recognized by farmers for centuries. It has been claimed by some that in 1660 farmers in the vicinity of Rouen, France, secured passage of a law which required destruction of the barberry bushes in that area.

The authenticity of this report has never been firmly established. We are certain, however, that colonists took the barberry to New England with them and that wheat rust became increasingly destructive after 1660. A law was passed in Connecticut in 1726 which authorized town meetings to adopt barberrye radication measures. Similar legislation was enacted in Massachusetts in 1755 and in Rhode Island in 1766. It should be emphasized that in the development of the control practices mentioned above there is no evidence that anyone had the present-day concept that many diseases are brought about by separate infective entities under the influence of favourable environment.

Theophrastus recorded his observations on the effects of various diseases of plants and in the main assumed that the basic causes lay in the vicissitudes of the weather. Pliny the Elder wrote upon the diseases of trees and of the cereal grains, but his work is little more than a translation of that of Theophrastus. Giambattista della Porta's discussion of diseases of plants, consisted largely of quotations from Theophrastus and Pliy, with some observations of his own. In 1705, the" French systematist Tournefort classified plant diseases, dividing them into two major groups, one due to external causes, the other to internal causes.

There followed a period of 150 years in which most writings on plant diseases consisted in descriptions of their appearances or symptoms and an arrangement of them into classes, genera, and species. The influence of the Linnaean binomial system was- evident in each of these compilations. Adanson in his work on "Familles des plantes" enumerated 23 species of plant disease, which were divided into two classes unpending upon whether they were due to internal or external causes. The nature of the causes was however, primarily a matter of speculation.

In Rozier's "*Cour complet dagriculture*" the list was expanded to 2 32 species. Plenck divided diseases into eight classes: external injuries, flux of juices, debility, cachexias, putrefactions, excrescences, monstrosities, I sterilities. In 1773, Zallinger placed plant diseases in five classes: phlegmasiae, paralyses or debilities, "cudischarges and drainings, cachexia, or bad constitution of the body, chief defects of different organs. Fabricius, a Danish botanist who student. Wied under Linnaeusand was for a time on the staff of the Royal 'Bot anical Garden in Copenhagen, published a classification of plant diseases theses in 1774. He spent much time in Paris, and evidently drew much of an his material from Adanson's book. His treatise is an arrangement of diseases into classes, genera, and species on the basis of their symptoms. 1: of That Fabriciu as not satisfied with current explanations of certain plant diseases is indicated by his statement that "certain It is that the of cases and symptoms of smut can never 11 better explained than by as summing something organized to be the cause.

In 1755, Tillet director of the mint at Troyes, France, and amateur botanist and agriculturalist, published an important paper If on wheat bunt. He found that opinion varied widely as to the cause of 1 the-disease-evil fogs, winds, excessive soil moisture; improper plant nutrients, soil texture, insects. In well-planned field experiments he tested the effect of dusting seed with the black mass contained in diseased kernels, and of treating such seed' with water containing- common salt and Plime. He demonstrated that the disease was contagious, since dusted a seed produced a much larger percentage of diseased plants at maturity than did undusted seed, while seed treated with salt and lime produced a lower percentage of diseased plants.

Tilet did not recognize that the smut dust consisted of spores which were "seeds" of a parasitic fungus. He rather looked upon the spore dust as merely the carrier of a poisonous entity. However, he did establish for the first time that the disease was 2 in some way contagious through the smut dust and that healthy seed produced only healthy plants. Furthermore, he demonstrated clearly that chemical treatment inhibited the contagious entity. In 1759, Count Francesco Ginanni in Italy published a book on the same subject entitled *"Trattato delle malatie del grano in erbo"* in which he attributed the cause of bunt to minute Insects. Another Frenchman, Tessler, working in the same period, published a paper entitled "Traite des maladies, It des' grains," in which he attributed rust to a stoppage of the transpiration 'I of the plant, caused by mists.

In 1766 an Italian, Giovanni Targioni Tozzetti, put forward al theory that the rusts and smuts of wheat consisted' of the vegetation of fungi arising under the epidermis of the plant. Fontana an Italian physicist, published a paper the following year in which he expressed the view that the grain rust was a distinct parasitic entity.

This new theory received some support, no doubt, but the weight of opinion was against it. In this same decade Anderson, an Englishman, wrote concerning the degeneration of potato varieties which we now know to be due to infectious viruses. In 1778 he published evidence to show that the malady was tuber-borne and likened it to smallpox in its infectiousness.

Zallinger in his book on plant diseases, "Demorbis plantarum" upheld the prevalent concept that fungi found in association with plant-disease lesions were abnormal structures of the diseased plants.

MODERN PLANT PATHOLOGY

It is apparent that at the beginning of the nineteenth century when Persoon and, a little later, Fries were actively naming and compiling fungi without reference to their relation to plant disease, those who were writing about these maladies were still not cognizant of the true relation to them of microorganisms. The first investigator to bring forth the facts which clearly demonstrated the pathogenic nature of any microorganism was Prevost, a Swiss professor of philosophy at the Academy at Montauban, France. He studied the wheat-bunt disease Jor about 10 years beginning about 1797. He published in 1807 his epochal "Memoir on the immediate Cause of Bunt or Smut of Wheat, and of Several Other Diseases of Plants, and on Preventives of Bunt". After presenting an accurate description of the disease, he reported important results of experiments upon it.

He studied and described the germination of the spores. He demonstrated, in confirmation of Tillet, that spores applied to the seed in some way resulted in infecting the plant. He studied the growth of the- fungus within the growing wheat plant and its final development in the embryo where spores were formed, and where the life cycle of the fungus was completed. He showed that solutions containing copper sulfate prevented spore germination. He distinguished between fungicidal and fungistatic effects. From these facts he demonstrated the basis of control through seed treatment, a practice which had been carried on empirically and without any rational explanation for 100 years and more.

The thorough hold of the theory of spontaneous generation on men's minds at this time was demonstrated by the fact that Prevost's results and interpretations were not accepted. A committee of the French Academy rejected his conclusions as unsound, and some 40 years had to elapse before others confirmed Prevost and pressed the adoption of the germ theory.

"In the period from 1807 to 1845 the prevalent teachings are illustrated in the writings of Count: Fillipo Re, Franz Unger and Franz-Julius Ferdinand Meyen. Re was professor of botany and agriculture at the University of Modena, Italy. His treatise on the diseases of plants was published in 1807, in the same year as the publication of Prevost's memoir. A second edition was published in 817, and, in 1849 Berkeley published an English translation of the latter in the Ci Gardener 8 Chronicle, stating that it was the best book available on the subject. It is of interest to take more than casual note of Re's book, because it portrays the increasing conflict of thought on the nature of plant disease in the beginning of the nineteenth century. Diseases were divided into classes and genera on the basis of their symptoms and their a supposed causal factors. The plan was not fundamentally different C from that of classifications Tesented in the last half of the previous century. The influence of current writings on the classification of human, diseases is evident. We need mention here only the titles of the five classes: diseases constantly sthenic; diseases constantly asthenic; diseases derived, sometimes from the sthenic, sometimes from the asthenic state; mechanical injuries, wounds; indeterminate diseases. In the last class were those diseases which author: regarded as of uncertain origin.

The pressure of the parasitic theory is evident from the remarks of Re who evidently did not see the work of Prevost before He second edition of his book was Published in 1817. The experimental evidence was racking, except that presented by Prevost, to clinch the germ theory. The adherents of the autogenetic theory of disease still had the better of the argument.

Franz Unger was born in Germany but spent most of his life in Austria Hungary. He was trained in medicine but in early mid-life turned to botany, becoming a professor of botany in Graz in 1835 and. professor of vegetable physiology in Vienna in 1849. In 1833, while serving as a medical officer in the Tirol, he published a book entitled "Die Exantheme 2 der Pflanzen". He believed that the fungus organisms associated with disease were endophytes but not parasites, that, they had no independent existence, and that under certain environmental conditions every plant had the potentiality of producing them, in some way, from the waste material or exanthema of the plant. The influence of human pathology upon the interpretation of plant disease was still strong.

Meyen, a professor of botany in Berlin, was concerned primarily with It' plant anatomy and physiology, but he wrote a look in plant pathology I which was published in 1841. In the year following his death. He followed Tournefort's system of classification of plant diseases. He considered smuts of cereals to be non-infections and to be due to a rotting of to sap brought about by Improper fertilization of the oils Rusts and mildews 4, were likewise regarded as the by-products of disease due to environmental factors. He described the formation of the pustules of white rust o shepherd's-purse in a unique manner.

He claimed that as the disease attacked the plant, cells of the latter enlarged and stretched out in some way towards the epidermis and cut off several spores in succession. Thus this incomplete picture of the development of sporangia as we now know it was misinterpreted to strengthen the autogenetic theory of disease.

About 1830, a few decades after the culture of potato as a food crop had expanded rapidly in Europe and North America, the late-blight disease appeared in both continents. It reached epidemic proportions by 1845. In that year it received much attention by writers in agricultural papers in Europe and the United States. The editorials and comments by contributors in the Garaener's Chronicle of 1845 are particularly enlightening to the student of plant pathology because they reveal better than any other existing evidence the trend of thought of laymen and scientists on the nature and cause of this destructive disease. While the association of the causal fungus with the blight of foliage and the rot of tubers was recognized by all who studied the disease closely, most writers held, in., accord with the autogenetic theory of disease upheld by Unger and by Meyin, that the fungus was the result rather than the cause of the disease.

Von Martius, in Germany, is one of the first to distinguish the fungus. 'He ascribed the disease to the effects of the fungus and published drawings of the latter. In 1845, Morren, director of the school of agriculture at Liege, Belgium, where the disease had dovastated the potato fields, described the disease and the associated with it He conducted crude inoculation experiments with the fungus and concluded that it was the major cause. He suggested the use of a mixture of lime, salt, and copper sulfate in suspension in water as a drench of the soil to reduce tuber rot.

Had he applied this mixture to the foliage, he might have anticipated the discovery of Bordeaux mixture by 40 years. In the same year, Montagne, a French mycologist, collected the fungus in France, named it Botrytis infestans, and sent specimens to Berkeley in England. The spreading epidemic had not yet reached Berkeley's home in North amptonshire, but it soon did. He confirmed Montagne's diagnosis and description but hedged on dogmatically upholding the idea that the fungus was the primary cause of the- disease.

English horticulturists, led by Lindley, editor of the Gardener's Chronicle, maintained that the unusual cool moist weather had so upset the normal growth of the potato that an internal breakdown of sort followed. The fungus was universally present but entirely secondary, feeding upon and multiplying in the dead plant tissue. Lindley headed a commission to investigate the disease in Ireland, where acute distress and famine followed in the wake of the epidemic. After what Lindley and his associates thought to be a thorough investigation, the weather got the blame, and the fungus remained innocent, in spite of increasing evidence from some quarters, especially in America, that blight and rot occurred even where the weather was not abnormally cool or moist.

In 1846 Berkeley took the stand that Botrytis infestans was the primary cause of the blight and tuber rot of potato. He maintained, further, that wheat bunt was incited by a fungus. In 1848 he reported the potato fungus on Datura stramonium L., on Anthocercis, and on tomato. He stated again with regard to the case of the disease: "I have no more doubt that the malady is caused by the fungus than I have that many of the cereal diseases are caused by fungi."

What was lacking was experimental proof that be infestans was the primary cause of the disease. Morren had provided some evidence, but the techniques of those days were still not adequate to refute completely the interpretations of the autogeneticists. Undoubtedly the supporters of the germ theory were increasing in number. Its general acceptance awaited only more clear-cut experimental evidence.

In the decade of 1840 Louis Rene Tulasne and Gharles Tulasne, botanists at the Museum of Natural History in Paris, began extensive morphological studies of the fungi. Their excellently illustrated-accounts of rusts, smute, and Ascomycetes are classics in mycology. In 1847, they published studies on the wheat-bunt fungus which confirmed the observations of Prevost with reference to spore of germination and relation of the fungus to the' host plant. They extended m their observations to several other smut organisms and to several rust fungi. The work of the Tulasne brothers was followed in 1853 by a comprehensive paper on the fungi, including the smuts and rusts, by Anton De Bary. Trained in medicine, De Bary early: turned to m: botany as a career. This paper, published when he was 22 years of age, up was dedicated to Alexander Braun, his professor of botany at Berlin, and a student of the algae.

He studied the development of several fungi in a critical manner. He Taking first the corn smut, he traced the development of mycelium, sporiferous branches, and chlaydospores. He was impressed by the a; fact that the mycelium in the intercellular spaces was distinct from rather Litoan a part of the host cell wall. In fact he pointed out that Unger, had I published the same observation in 1840. As the spores matured, he, noted that the mycelium passed into the spores and the hyphae becalne indistinct.

He saw that as the cllnsmut gall began to develop, the host cells underwent continuous division up to about the time when spore formation was, initiated. Then cell division slowed down gradually, and eventually) the host tissue was broken down as the rapidly growing fungus drew 1 nourishment' from the cells. Three other smuts were studied, and the c conclusion was drawn that all four species followed the same process in their development and differed only in morphological characteristics. He studied critically the development of pustules of white rusts-of several hosts and again interpreted- the mycelium and sporogenous hyphae as I distinct from the host cell and not arising from the latter as maintained by Meyen.

De Bary examined critically several of the common rust fungi including that of black stem rust' of cereals and that of rust of bean. Here he did not distinguish between summer spores,' and winter spores but described the latter and their formation accurately. Dimorphism of these species- was not described until the next year by Tulasne. The stages of rusts were till considered as separate After Persoon, few workers opposed the recognition of the smut particles as fungi; however, this did not prevent people from regarding them as products of disease in plants or from attributing the latter to entirely different causes entities by my ologists, and De Bary was in agreement with that concept.

He stated that although the.opinion was universal that barberry was harmful to grain, there was of that the spares.of Aecidium on barberry caused infection of cereals. Although he later was the first to prove this to be the fact, he was quite skeptical of the possibility in. 1853.

Meyen had suggested that the pycnia which produced pycniospores in association with Aecidium, Roestelia, and other aecial rusts were male organs. De Bary discussed this possibility at length but was definite in his contention that no pro of had been offered. He expressed hope that the matter of the relation of pycnia to aecia would be settled promptly since several workers were investigating the subject. As a matter of history, the question was not settled until Craigie presented evidence far heterothallism and explained the function of pycniospores in the rusts in 1927.

Had De Bary publish, only the first two parts of his paper the publication would have stood merely as a substantial contribution to the growing information on the fungi and their life histories and to the tax on omic relations of the farms considered. As to the tax onomy presented in Part II, it is evident that De Bary's attempts at classification were far from final. In 'Part III, however,' he launched into the controversial field of the relation of fungi to plant disease.

Reviewing the opinions which had prevailed from the time of Theophrastus,' he made it clear that the fungi associated with the diseases known as smuts' and rusts were regarded by mast farmers and botanist of De Bary's time as the effects of there, than the causes of disease. 'This, he explained, was because they saw these fungi appear simultaneously with the degeneration or failure of the-hast plants and were unable to study the formation of these degenerative lesions Thus all manner of external factors were' blamed as-causes. As ideas changed other theories arose to explain the association of the organisms with disease. Linnaeus considered the Ostalgia spares as eggs of Infuse oria. Later the similarity of smut and rust particles to fungus spares led him to include them among his fungus genera. Finally Persoon introduced a definite classification in which the genera Aecidittm Uredo, and Puccinia were set up.

In brief outline we have scanned some of the major workers and their contributions in mycology and plant pathology up to the time of De Bary in 1853. We have emphasized that throughout the 18th century and the first half of the nineteenth century the taxonomy of plants was I paramount and the dogma of constancy of species had a strong influence upon the trend of botanical thought. Moreover, the theory of spontaneous generation of microscopic forms and of the fungi generally held sway. Parasitism was not an entirely new concept, for the parasitic phanerogams had been generally accepted as true parasites upon their hosts-before this I date, and De Bary in his first paper referred repeatedly to this fact in his I argument for the acceptance of fungi as plant parasites.

It is important that we realise that other phases of botany also passed although transition stages as did mycology. The approach to natural I science by experimentation and inductive" reasoning was also affecting plant physiology and morphology.

Robert look was a British scientist and philosopher who by 1660 had greatly improved the compound microscope. He noted the cellular structure in charcoal and in cork. He published in 1667, a book under the title "Micrographia or Some Physiological Descriptions of Minute Bodies Made by Magnifying Glasses". Two contemporaries of Hooke were Malpighi, Italian physician and anatomist, and Grew, British physician, who pursued studies of plant anatomy as an avocation. These two men presented results of their studies before the Royal Society of London in 1671, and both continued to publish in later years. Their work was largely exploratory and dealt with grosser features of anatomy. Malpighi published "Anatome plantarum" in 1675. Grew published "The Anatomy of Plants" in 1682. The important point of interest about these two men is that they advanced plant anatomy to a point which was not again exceeded for over 100 years.

Anton van Leeuwenhoek was contemporary with Malpighi and Grew. Not highly educated, he assiduously devoted his spare time to the preparation of superior lenses and made available to himself equipment capable of higher magnification than that employed by Ma1pighi and Grew. Born in Holland, where he carried on his work, he communicated his reports to the Royal Society of London. Besides discovering bacteria he also studied plant anatomy. His work as a whole was not so thorough as the studies of Malpighi and Grew, but with superior equipment he discerned some things which they overlooked, such as the crystals within certain plant cells.

Stephen Hales was the outstanding figure in plant physiology of the eighteenth century. His most quoted contribution is his book entitled "Vegetable Staticks", which was published in 1724. Also in the eighteenth century were Kohlreuter, early plant hybridities, and the Dutch physician, Ingenuous who pioneered in the study of photosynthesis. He was the first to conceive that green plants took up carbon dioxide from the air and gave of oxygen. His work was con temporary with and helped by that of the French chemist, Lavoisier, who discovered oxygen and carbon dioxide.

In the first half of the nineteenth century, De Saussure Swiss physiologist, advanced knowledge of the oxygen-carbon dioxide relation to plant nutrition and of the mineral nutrition of plants, while Dutrochet, also Swiss, studied the phenomenon of miosis. Organic chemistry became a distinct science with the synthesis of urea in 1828. But it should beet in mind that the advances of pioneers- 'just mentioned were not very generally accepted. The pull to revert to deductive reasoning and to glmophical rather than the experimental scientific approach was still strong.

Following De Saussure, Liebig became the dominant leader in plant nutrition and championed the us of mineral fertilizers in plant production. Nageli, German physical chemist and plant morphologist, was one of the great German minds of his period. He pioneered the field of physiology of lower organisms.

During the first half of the nineteenth century there was revival of interest in anatomy. It was in this period that Schleiden and the zoologist Schwann formulated their cell theory. In the field of plant anatomy von Mohl, Meyen, Unger, and Nageli were all active. Robert Browne is generally credited with the first observation of the cell nucleus. I Dujarden, French.zoologist, discovered the material within the cell we now refer to as protoplasm. Purkinje, Bohemian biologist, was; the first to use the term "ptoplasm".

PATHOGENIC FUNGI

Pathogenic fungi are fungi that cause disease in humans or other organisms. The study of pathogenic fungi is referred to as medical mycology. Although fungi are eukaryotic organisms many pathogenic fungi are also microorganisms.

Candida

Candida species are important human pathogens that are best known for causing opportunist infections in immunocompromised hosts (e.g. transplant patients, AIDS sufferers, cancer patients). Infections are difficult to treat and can be very serious: 30-40% of systemic infections result in death. The sequencing of the genome of *C. albicans* and those of several other medically-relevant *Candida* species has provided a major impetus for *Candida* comparative and functional genomic analyses. These studies are aiding the development of sensitive diagnostic strategies and novel antifungal therapies.

Aspergillus

Aerosolized *Aspergillus* spores are found nearly everywhere so we are routinely and almost constantly exposed to them. Such exposure is a normal part of the human condition and generally poses no adverse health effects. Nevertheless, *Aspergillus* can and does cause disease in three major ways: through the production of mycotoxins; through induction of allergenic responses; and through localized or systemic infections.

With the latter two categories, the immune status of the host is pivotal. Allergies and asthma are thought to be caused by an active host immune response against the presence of fungal spores or hyphae. In contrast, with invasive aspergillosis, the immune system has collapsed and little or no defence can be mounted.

The most common pathogenic species are *Aspergillus fumigatus* and *Aspergillus flavus*. Aspergillus flavus produces aflatoxin which is both a toxin and a carcinogen and which can potentially contaminate foods such as nuts.

Aspergillus fumigatus and *Aspergillus clavatus* can cause allergic disease. Some *Aspergillus* species cause disease on grain crops, especially maize, and synthesize mycotoxins including aflatoxin. Aspergillosis is the group of diseases caused by *Aspergillus*. The symptoms include fever, cough, chest pain or breathlessness. Usually, only patients with weakened immune systems or with other lung conditions are susceptible.

Cryptococcus

Cryptococcus neoformans can cause a severe form of meningitis and meningo-encephalitis in patients with HIV infection and AIDS. The majority of *Cryptococcus* species live in the soil and do not cause disease in humans. *Cryptococcus neoformans* is the major human and animal pathogen. *Cryptococcus laurentii* and *Cryptococcus albidus* have been known to occasionally cause moderate-to-severe disease in human patients with compromised immunity. *Cryptococcus gattii* is endemic to tropical parts of the continent of Africa and Australia and can cause disease in non-immunocompromised people.

Histoplasma

Histoplasma capsulatum can cause histoplasmosis in humans, dogs and cats. The fungus is most prevalent in the Americas, India and southeastern Asia. It is endemic in certain areas of the United States. Infection is usually due to inhaling contaminated air.

Pneumocystis

Pneumocystis jirovecii (or Pneumocystis carinii) can cause a form of pneumonia in people with weakened immune systems, such as premature children, the elderly, and AIDS patients.

Stachybotrys

Stachybotrys chartarum or "black mold" can cause respiratory damage and severe headaches. It frequently occurs in houses in regions that are chronically damp.

DRUG RESISTANCE

Treatment with antifungal drugs often results in the appearance of resistant strains of fungi.

Various mechanisms leading to resistance have been described.

For example, a number of resistant clinical isolates overexpress genes encoding drug efflux pumps.

Recent advances in molecular biology have allowed the study of the phenomenon of multi-drug resistance on a genome-wide scale. DNA microarrays are being used to study the expression profiling of pathogenic fungi and proteomics is aiding research in the development of resistance to various antifungal drugs.

Endothermy

Mammalian endothermy and homeothermy are potent nonspecific defenses against most fungi.

BASIC BIOLOGY OF FUNGI

Yeasts and Molds

These fungi grow as saprophytes, parasites, or both by using specific proteolytic, glycolytic, or lipolytic enzymes to extracellularly break down substrates and to absorb the products of digestion through the fungal cell envelope.

Cell Wall

The fungal cell wall gives shape and form, protects against mechanical injury, prevents osmotic lysis, and provides passive protection against the ingress of potentially harmful macromolecules.

Filamentous Fungi and Filamentous Bacteria

Fungi are different from the Actinomycetes, a group of prokaryotic filamentous bacteria having peptidoglycans in their cell walls and an absence of nuclear membranes and organelles, but the two groups of microorganisms are usually considered together in texts.

Hyphal and Yeast Morphogenesis

Hyphal extension growth occurs apically by a sophisticated organization of tip-growth-related organelles and cytoskeletal elements.

Hyphal wall and yeast cell wall polysaccharide synthetases are active at sites where growth is occurring and inactive when no growth is occurring. Morphogenesis is a balance between wall synthesis and wall lysis.

Sexual Reproduction

Sexual reproduction occurs by the fusion of two haploid nuclei (karyogamy), followed by meiotic division of the diploid nucleus.

The union of two hyphal protoplasts (plasmogamy) may be followed immediately by karyogamy, or it may be separated in time.

Asexual Reproduction

Asexual reproduction occurs via division of nuclei by mitosis. With the absence of meiosis, other mechanisms associated with the nuclear cycle result in recombination of hereditary properties and genetic variation.

Macroscopic fungi such as morels, mushrooms, puffballs, and the cultivated agarics available in grocery stores represent only a small fraction of the diversity in the kingdom Fungi. The molds, for example, are a large group of microscopic fungi that include many of the economically important plant parasites, allergenic species, and opportunistic pathogens of humans and other animals. They are characterized by filamentous, vegetative cells called hyphae.

A mass of hyphae forms the thallus (vegetative body) of the fungus, composed of mycelium. The more phylogenetically primitive molds (e.g., water molds, bread molds, and other sporangialsaclikeforms) produce cenocytic filaments (multinucleate cells without cross-walls), while the more advanced forms produce hyphae with cross-walls (septa) that subdivide the filament into uninucleate and multinucleate compartments.

The septum, however, still provides for cytoplasmic communication, including intercellular migration of nuclei. Many fungi occur not as hyphae but as unicellular forms called yeasts, which reproduce vegetatively by budding. Some of the opportunistic fungal pathogens of humans are dimorphic, growing as a mycelium in nature and as a vegetatively reproducing yeast in the body. Candida is an example of such a dimorphic fungus. It can undergo rapid transformation from the yeast to the hyphal phase in vivo, which partly contributes to its success in invading host tissue.

The true fungi obtain their carbon compounds from nonliving organic substrates (saprophytes) or living organic material (parasites) by absorption of nutrients through their cell wall. Small molecules (e.g., simple sugars and amino acids) accumulate in a watery film surrounding the hyphae or yeast and simply diffuse through the cell wall. Macromolecules and insoluble polymers (e.g., proteins, glycogen, starch, and cellulose), on the other hand, must undergo preliminary digestion before they can be absorbed by the fungal cell. This process involves release of specific proteolytic, glycolytic, or lipolytic enzymes from the hypha or yeast, extracellular breakdown of the substrate(s), and diffusion of the products of digestion through the fungal cell envelope. Fungal pathogens rely on these digestive enzymes to penetrate natural host barriers.

Cell Walls

Not all species of fungi have cell walls, but in those that do, cell wall synthesis is an important factor in determining the final morphology of fungal elements. Thus, our knowledge of fungal morphogenesis has evolved in parallel with our understanding of fungal cell wall biosynthesis. The fungal wall also protects cells against mechanical injury and blocks the ingress of toxic macromolecules. This filtering effect may be especially important in protecting fungal pathogens against certain fungicidal products of the host.

The fungal cell wall is also essential to prevent osmotic lysis. Even a small lesion in the cell wall can result in extrusion of cytoplasm as a result of the internal (turgor) pressure of the protoplast. The composition of fungal cell walls is relatively simple and includes substances not typically found in animal and plant hosts (e.g., chitin). On this basis, it may be possible to identify pathogen-specific molecular targets from investigations of the biosynthesis of fungal wall components. Such targets may prove pivotal for the successful development of antifungal drugs that are not toxic to mammalian cells.

FILAMENTOUS FUNGI AND FILAMENTOUS BACTERIA

Fungi, like bacteria, are ecologically important as decomposers as well as parasites of plants and animals. Both groups of microbes often inhabit the same ecosystem and thus compete for the same food supply.

Associated with this competition is the production by both the fungi and bacteria of secondary products that function as microbial growth inhibitors or toxins. These compounds constitute a rich library of antimicrobial agents, many of which have been developed as pharmacologic antibiotics (e.g., penicillin from *Penicillium chrysogenum,* nystatin from *Streptomyces noursei,* amphotericin B from *S niveus*).

The superficial morphologic similarities between actinomycetes (filamentous bacteria) and molds suggest that the two groups have undergone parallel evolution. Despite the production of branching filaments and mold-like spores, the actinomycetes are clearly prokaryotes, whereas fungi are eukaryotes.

Moreover, the sexual reproduction of bacteria, which typically occurs by transverse binary fission, should not be confused with asexual processes of budding and fragmentation associated with mitotic nuclear division in fungi. Most of the molds that produce septate vegetative hyphae reproduce exclusively by asexual means, giving rise to airborne propagules called conidia. On the other hand, elaborate mechanisms of sexual reproduction are also demonstrated by members of the Eumycota. Four distinct kinds of meiospores (products of karyogamy-meiosis-cytokinesis) are recognized: oospores (Oomycetes), zygospores (Zygomycetes), ascospores (Ascomycetes), and basidiospores (Basidiomycetes).

Hyphal and Yeast Morphogenesis

Hyphal growth occurs by extension at the tips. This polarization is at least partially determined by directional movement and accumulation of vesicles that carry wall precursors and wall synthetases to the site of exocytosis at the apical dome of the hypha.

Despite the apparent simplicity of hyphal morphogenesis, ultrastructural investigations have shown a sophisticated organization of tip-growth-related organelles and cytoskeletal elements. There is evidence that intussusception and polymerization of chitin microfibrils occur at the apical dome of the hypha and that the biosynthesis of this major cell wall product is controlled by the activity of membrane-bound chitin synthetase. The zymogen form of chitin synthetase has been detected in microvesicles called chitosomes, which appear to transport this enzyme to the hyphal tip. The chitosomes may arise from Golgi-like bodies or by a process of self-assembly of subunits freely within the cytoplasm or within larger vesicular bodies.

Activation of chitin synthetase occurs upon fusion of the chitosome with the plasmalemma and may be due to the interaction of a membrane-bound protease and the zymogen. Chitin microfibrillogenesis is initiated at these sites of fusion. Evidence has also been presented, primarily from studies of the yeast *Saccharomyces cerevisiae*, that biosynthesis of skeletal polysaccharides is catalyzed by polysaccharide synthetases (e.g., chitin synthetase and b1-3-glucan synthetase), which are uniformly distributed within the plasmalemma. These wall-synthesizing, cell-membrane-bound enzymes occur in either zymogen or active forms. The model of yeast morphogenesis suggests that the synthetase is active at sites where the wall is growing and inactive where it is quiescent. One possibility is that microvesicles transport activating factors (e.g., proteases, ATP, and GTP) to the plasmalemma at specific sites of wall biosynthesis (zones of bud emergence and of septal formation). These two concepts of regulation of wall biosynthesis in fungal hyphae and yeasts have been supported by considerable bodies of evidence, and it is likely that both are correct.

Extension growth of hyphal tips and yeast buds logically requires a balance between processes of insertion of newly synthesized polymeric material and modification of the existing microfibrillar matrix to accommodate expansion and further intussusception of wall polymers. In other words, a balance between wall synthesis and wall lysis, or plasticization, is essential for maintaining the orderly processes of hyphal tip elongation and bud emergence.

The presence of lytic enzymes in the fungal wall has been reported, including, b1-3-glucanase, N-acetyl-b-D-glucosaminadase, and chitinase. Localization of such activity may be mediated by macrovesicles. These organelles, like microvesicles, are probably derived from Golg-like bodies and are directed to the hyphal tip or yeast bud and fuse with the plasmalemma, thereby delivering their contents to the site of wall synthesis.

Reproduction

Sexual reproduction in the fungi typically involves fusion of two haploid nuclei (karyogamy), followed by meiotic division of the resulting diploid nucleus. In some cases, sexual spores are produced only by fusion of two nuclei of different mating types, which necessitates prior conjugation of different thalli. This condition of sexual reproduction is known as heterothallism, and the nuclear fusion is referred to as heterokaryosis.

Normally plasmogamy (union of two hyphal protoplasts which brings the nuclei close together in the same cell) is followed almost immediately by karyogamy. In certain members of the Basidiomycotina, however, these two processes are separated in time and space, with plasmogamy resulting in a pair of nuclei (dikaryon) contained within a single cell.

Karyogamy may be delayed until considerably later in the life history of the fungus. Meanwhile, growth and cell division of the binucleate cell occur. The development of a dikaryotic mycelium results from simultaneous division of the two closely associated nuclei and separation of the sister nuclei into two daughter cells.

An alternative mechanism of sexual reproduction in the fungi is homothallism, in which a nucleus within the same thallus can fuse with another nucleus of that thallus (i.e., homokaryosis). An understanding of these nuclear cycles is fundamental to investigations of fungal genetics. As mentioned above some fungi are classified as strictly asexually reproducing forms. These include the large group of asexual (imperfect) yeasts (e.g., *Candida* species) and conidial fungi (e.g., *Coccidioides immitis*). Most members of this group have permanently lost their ability to produce meiospores. A few undergo rare sexual reproduction, and perhaps for some species we have yet to discover their sexual (perfect) stage.

The most common methods of asexual reproduction, in addition to simple budding in yeasts, are blastic development of conidia from specialized hyphae (conidiogenous cells), fragmentation of hyphae into conidia, and conversion of hyphal elements into conidia or chlamydospores (thick-walled resting spores).

Despite the absence of meiosis during the life cycle of these imperfect fungi, recombination of hereditary properties and genetic variation still occur by a mechanism called parasexuality. The major events of this process include the production of diploid nuclei in a heterokaryotic, haploid mycelium that results from plasmogamy and karyogamy; multiplication of the diploid along with haploid nuclei in the heterokaryotic mycelium; sorting out of a diploid homokaryon; segregation and recombination by crossing over at mitosis; and haploidization of the diploid nuclei. Sexual and parasexual cycles are not mutually exclusive. Some fungi that reproduce sexually also exhibit parasexuality.

An extensive foundation of knowledge on the basic biology of fungi is at hand, including fungi that cause superficial, deep-seated, and systemic infections of humans and other animals. Much less is known, however, of the intricacies of interactions between these largely opportunistic pathogens and their hosts. Many areas of research in medical mycology are still in their infancy and offer formidable challenges and potential rewards. The current application of methods of recombinant DNA technology to problems of fungus-host interactions, especially the identification of pathogenicity genes, holds promise for significant contributions to our knowledge of medically important fungi.

6

Control of Plant Disease

Decay fungi are living organisms which send minute threads called "hyphae" through damp wood, taking their food from the wood as they grow. Gradually, the wood is decomposed and its strength is lost. Such damage is often inconspicuous until its final stages, and in a few instances homeowners have suddenly found floors breaking through or doors falling from their hinges due to wood rot. When previously dry wood is placed in contact with moist soil, or in a location where it is subject to condensation (such as unventilated crawl space), it is likely that wood decay problems will occur.

Rain leaks, faulty plumbing and leaky downspouts also are common sources of moisture. In some instances, water can be transported to the site of decay through strands or "rhizomorphs" of the decay fungi. Water-transporting strands may extend for thirty or more feet across brick, concrete or similar materials.

The wood decay fungus, *Serpula lacrimans*, has been known to transport water up three stories to an area where decay is occurring. *Poria incrassata* is also capable of transporting water long distances. However, these fungi are exceptions to the rule. Most wood-rotting fungi must have a direct supply of water at the site of decay. Thus the term "dry-rot," sometimes applied to decay in wood structures, is erroneous.

PREVENTION

- If the decay hazard is high, select the heartwood of decay-resistant species or use wood properly treated with a good preservative. Conifers from which decay-resistant lumber is produced include Pacific yew, juniper, redwood, baldcypress, and western red cedar. Durable hardwood species include osage orange, black locust, red mulberry, catalpa and black walnut.

- Build on a well-drained site. Use proper grading to prevent water from seeping under the house. Install effective drain tile, roof overhang, gutters, and downspouts. Place no untreated wood within 18 inches of the ground.
- Provide adequate cross ventilation beneath buildings to eliminate dead air pockets. Install two square feet of opening for 25 linear feet of wall. Dense bushes or other plants should not be placed in front of these ventilators.
- Install a vapour barrier on the soil surface to cause soil moisture to condense on the barrier and return to the soil rather than condensing on the floor and above joists. Satisfactory barriers can be made by covering the soil with asphalt roofing paper or polyethylene sheets.

REPAIR OF DECAYED BUILDINGS

First determine the source of moisture and remove it. If adequate ventilation and soil drainage are provided and all contacts of untreated wood with the soil or moist concrete or masonry are broken, decayed wood will dry out and further decay will be stopped. When making replacements, cut out at least one foot beyond the rotten area. Avoid placing new lumber in contact with old, decayed wood. Replacement lumber should be treated before installation. Remodel to provide more ventilation and better design rather than simply replacing decayed lumber.

Bacterial Crown Gall of Fruit Crops

Crown gall is caused by the bacterium *Agrobacterium tumefaciens.* This bacterium has the widest host range of any plant pathogen. It is capable of causing tumors, or "galls," on virtually all plant species, except the monocots (grasses). A similar bacterium, *Agrobacterium rubi,* causes galls on the canes of brambles. All fruit crops grown are susceptible.

The disease is particularly destructive on brambles (raspberries and blackberries) and grapes. It can also cause severe problems on apple, pear, blueberry, all stone fruits and on ornamentals. The bacteria induce galls or tumors on the roots, crowns, trunks and canes of infected plants. These galls interfere with water and nutrient flow in the plants. Seriously infected plants may become weakened, stunted and unproductive.

Symptoms

The disease first appears as small overgrowths or galls on the roots, crown, trunk or canes. Galls usually develop on the crown or trunk of the plant near the soil line or underground on the roots. Above ground or aerial galls may form on canes of brambles and highly susceptible cultivars of grape. Although they can occur, aerial galls are not common on fruit trees. In early stages of development the galls appear as tumor-like swellings that are more or less spherical, white or flesh-coloured, rough, spongy (soft) and wart-like. They usually form in late spring or early summer and can be formed each season.

As galls age they become dark brown to black, hard, rough, and woody. Some disintegrate with time and others may remain for the life of the plant. The tops of infected plants may appear normal. If infection is severe, plants may be stunted, produce dry, poorly-developed fruit, or show various deficiency symptoms due to impaired uptake and transport of nutrients and water.

Causal Organism

The crown gall bacterium is soil-borne and persists for long periods of time in the soil in plant debris. It requires a fresh wound in order to infect and initiate gall formation. Wounds that commonly serve as infection sites are those made during pruning, machinery operations, freezing injury, growth cracks, soil insects and any other factor that causes injury to plant tissues.

Bacteria are abundant in the outer portions of primary galls, which is often sloughed off into the soil. In addition to primary galls, secondary galls may also form around other wounds and on other portions of the plant in the absence of the bacterium. The bacteria overwinter inside the plant (systemically) in galls, or in the soil. When they come in contact with wounded tissue of a susceptible host, they enter the plant and induce gall formation, thus completing the disease cycle. The bacteria are most commonly introduced into a planting site on or in planting material.

Control

- Obtain clean (disease free) nursery stock from a reputable nursery and inspect the roots and crowns yourself to make sure they are free from galls. Avoid planting clean material in sites previously infested with the bacteria.
- Avoid all unnecessary root, crown and trunk wounding by careless cultivation and other machinery operation, and control soil insects. Any practice that reduces wounding is highly beneficial. Preventing winter injury (especially on grapes) is also beneficial.
- On grapes, the double trunk system of training may be a useful system for minimizing losses due to crown gall. If one trunk is infected, it can

be removed. The remaining trunk can be pruned leaving a full number of buds until the second trunk can be renewed. Galls on the upper parts of the trunk or on canes can be removed by pruning.

- A relatively new biological control agent for crown gall is available for apple, pear, stone fruit, blueberry, brambles and many ornamentals. It is not effective on grape. The agent is a nonpathogenic strain of bacterium (*Agrobacterium radiobacter* strain 84) that protects the plants against infection by the naturally occurring strains of pathogenic bacteria in the soil. Nursery stock is dipped in a suspension of commercially prepared *Agrobacterium radiobacter* strain 84 at planting time. The antagonistic bacteria act only to protect disease free plants from future infection by the crown gall bacterium; they cannot cure infected plants.

BITTER ROT OF APPLES

Bitter rot is a common disease of apples and pears in practically all countries where they are commercially grown. Of the three fruit rot diseases on apple (bitter rot, white rot and black rot), bitter rot has the potential to be the most destructive. The fungus that causes bitter rot (fruit rot) can also cause a leaf spot and canker, although the leaf spot and canker form of the disease are not common. The disease is most common in warmer regions because high temperatures favour disease development. Before the development of effective fungicides, entire crops were lost to bitter rot during periods of warm, wet weather.

Symptoms

Fruit rot symptoms differ somewhat, depending on whether infection is initiated by spores from perithecial strains of the fungus (strains that produce ascospores and conidia) or conidial strains of the fungus (strains that produce only conidia). Initial symptoms produced by perithecial or conidial strains are similar. Lesions begin as small, slightly sunken areas, which are light brown to dark brown.

On mature fruit, lesions may be surrounded by a red halo. Lesions originating from infections by conidial strains remain circular and become sunken as they enlarge. Copious quantities of ooze containing conidia (spores) are produced in fungal fruiting bodies called acervuli, which occur in concentric circles on rotted tissue around the point of infection.

Acervuli are sparse on some lesions and very dense on others. Under moist, humid conditions, the spore masses appear creamy and are salmon to pink in colour.

Lesions initiated by perithecial strains are usually not sunken and are often darker brown than those caused by conidial strains. Acervuli are widely scattered over the surface, and perithecia are found in dark brown to black clumps scattered on the surface.

Bitter rot lesions on all fruit (regardless of which strain caused infection) extend in a cone shape manner toward the core. In cross section, the lesion appears "t-shaped". This is a reasonably reliable characteristic that can be used to separate bitter rot from white rot or black rot.

The rotten area is brown but much firmer than white rot. Infected fruit eventually mummify, and some may remain attached to the tree through the winter. Leaf lesions are not common and are caused by the perithecial strain. They begin as small, red flecks, which enlarge to irregular brown spots 1/16 to 1/2 inch in diameter. Severely affected leaves may drop off the tree. Bitter rot cankers are rare in the eastern United States. Cankers are oval, sunken and often zonate in appearance.

Casual Organism and Disease Cycle

Bitter rot is caused by the fungus, *Glomerella cingulata*. The fungus overwinters in apple orchards in dead wood or mummified fruit in the trees that were infected during the previous season. Conidia, produced in these overwintering sites, are the primary inoculum source in the spring, although ascospore inoculum is important in some orchards. Conidia are spread by splashing and wind-blown rain. Insects and birds are also involved in their dispersal.

Ascospores are released after rain and are airborne. Fruit are susceptible to infection from 3 weeks after petal fall until harvest. Temperatures of 80 or 90 degrees F are most favourable for disease development. Because of the large number of conidia produced in lesions on fruit, the fungus has the potential for rapid spread within the orchard.

Control

- Control of bitter rot is best achieved through an integrated programme of cultural practices and chemical control measures.
- Sanitation is critical for effective control. Piles of prunings are an important source of inoculum and should be removed from the perimeter of the orchard or burned. Prunings can be left on the orchard floor if they are chopped with a flail mower, which removes much of the bark and allows them to decompose faster.

  Removal of mummified apples and pruning out dead wood in the tree are important for reducing the inoculum within the tree. Pruning out current-season shoots infected with fire blight is also important, because they can be colonized and serve as an inoculum source during the same growing season.

- Any practice that helps to maintain trees in a healthy vigorous condition is critical for controlling the canker phase of the disease. Cankers generally develop only on stressed or weakened trees. Prune trees annually and maintain a balanced fertility programme based on soil and foliar nutrient analysis. Cankers generally develop rapidly on winter-injured trees.
- The use of fungicides combined with good sanitation is beneficial for controlling the fruit rot phase of the disease. Fungicides are not effective for controlling the canker phase of the disease on weakened trees.

WILD MUSHROOMS

There are 2,000 or more kinds of wild mushrooms. Some are poisonous and some are edible and delicious when properly prepared. The edibility of the majority is either not known or they are not considered for food because of their small size or poor flavour or texture.

Even though not every one is interested in collecting mushrooms to eat, it is important to understand most have an important and beneficial role in the environment. They grow in a wide variety of habitats. Most of the mushrooms seen on a walk through a woods are beneficial. Many species are quite specific about their food source and will be found only under or near certain kinds of trees-some under pines, others under oak, etc. Some are important as decay organisms, aiding in the breakdown of logs, leaves, stems and other organic debris.

This important role of mushrooms results in recycling of essential nutrients. Some mushrooms grow in, and form their fruiting structures on living trees causing decay of the sapwood or of the heartwood. Many woodland mushrooms are essential to good growth, and even survival of trees. They establish a relationship with roots of living trees that is mutually beneficial. These are called mycorrhizal mushrooms.

All mushrooms, whether poisonous or edible can be admired for their beauty and the fantastic variety of form, colour and texture.

Which Mushrooms are Safe to Eat?

Some edible mushrooms are very similar in appearance to poisonous kinds and may grow in the same habitat. Edible mushrooms are known to be safe to eat because they have been eaten frequently with no ill effects. Poisonous mushrooms are known because someone ate them and became ill or died. There is no test or characteristic to distinguish edible from poisonous mushrooms. This indicates a need to identify with certainty one of several of the proven edible species and pick and eat only those positively identified.

At the same time, you should also learn to identify some of the common poisonous mushrooms, especially those that are similar to edible kinds. It is especially important to learn the characteristics of the Amanita mushrooms, since several of the species common are poisonous, a few causing serious illness and sometimes death.

The word "toadstool" is often used to indicate a poisonous mushroom. Since there is no way to distinguish between a so-called "toadstool" and an edible mushroom it is more precise to speak of poisonous mushrooms or edible mushrooms. The season for collecting wild mushrooms for food begins in late March and early April when the first morel or sponge mushrooms are found.

These choice edible mushrooms are most abundant during April and the first two weeks of May. The false morels (members of the Gyromitra genus) are found at this same time of the year, but they must be regarded as poisonous and not collected for eating. It is true that many have eaten false morels with no apparent ill effects. However, recent research has shown toxins to be present in some of the false morels that can cause death or serious illness. Do not eat the false morels.

From mid summer to late autumn, a great variety of mushrooms may be found. A number of these are choice edibles. Photographs and brief descriptions of several of the more common mushrooms found are included in this fact sheet.

Control of Nuisance and Detrimental Molds (Fungi) in Mulches and Composts

Mulches and composts are often used to improve soils and plant health and to control weeds. They improve drainage as they decompose even though the ability of the soil to hold moisture is increased. They lower soil temperature in the summer and insulate roots from cold in winter conditions. Eventually, they mineralize, release nutrients for plants, and leave humic substances as residues. Their beneficial side effects gradually disappear unless more mulch or compost is applied.

Generally, these organic materials inhibit undesirable microorganisms such as soilborne pathogens that cause diseases of plants. They also stimulate the activity of many types of beneficial microorganisms, including mycorrhizal fungi. Occasionally, however, microorganisms (primarily fungi) in mulches and composts can become a nuisance and even cause certain diseases of plants.

Whether a mulch or a compost provides beneficial or detrimental effects is largely determined by the type of organic matter from which it was produced and the degree to which it was decomposed and treated before its application in the landscape. The temperature, pH, and moisture content of the products just before application also have an effect. The severity of nuisance fungi can be minimized if appropriate steps are taken in time.

Examples of Nuisance Fungi

The shotgun or artillery fungus (*Sphaerobolus*) may cause serious problems. While it decays the mulch, it also produces fruiting structures that resemble tiny cream or orange-brown cups that hold a spore mass resembling a tiny black egg (1/10 inch in diameter). This fungus shoots these spore masses high into the air. They stick to any surface and resemble small tar spots on leaves of plants or the siding of homes. They are difficult to remove, leave stained surfaces, and may result in major damage.

Slime molds are another type of nuisance fungus. They first appear as bright yellow or orange slimy masses that may be several inches to a foot or more across.

They produce tiny spores that eventually dry out and blow away. These molds, like many others such as stink horns and bird[1]s nest fungi, actually should be considered microbial ornamentals in the landscape. However, some fungi in mulches and composts produce toad stools (mushrooms), and some of these are toxic to humans. It is a good idea to destroy them when small children have access to the mulched area.

Another fungal problem that is often not identified correctly occurs when mulches are applied too deep (4-6 inches) instead of the ideal depth of 1.5 to 2 inches. Deep layers of mulch, particularly if prepared from fresh woody materials, may actually undergo high temperature decomposition during the summer.

The result is that the mulch dries out to less than 34 per cent moisture and becomes a dusty mass. Fungi often colonize these dry mulches until they become a water-repelling, moldy chunk of material. Young trees mulched in this way sometimes die from drought even though the homeowner irrigates the area, because water runs off the mulch since it repels water.

Other Types of Problems

Fresh mulches prepared from trees killed by plant diseases may be colonized by plant pathogens. *Verticillium dahliae,* a fungus that causes wilts and death of many shade trees and ornamental shrubs, can be carried in infested mulch and kill susceptible plants in the landscape. *Rhizoctonia solani,* another plant pathogen that causes damping-off of many types of seedling plants, is actually stimulated by fresh mulches. This pathogen utilizes the cellulose in wood as a source of food.

Short-term composting of mulches in windrows under high-temperature conditions (130-160 degrees F) kills these plant pathogens. Six weeks of composting is sufficient to kill most plant pathogens and avoid their dissemination in mulches or composts.

Mycorrhizae, which are fungi that can form beneficial associations with roots, also are affected by mulches and composts. A shallow layer of wood chips (1-2") or compost improves tree establishment because mycorrhizae are stimulated by the slow release of organic sources of nitrogen and carbon in organic matter. However, a deep layer (4-6") of the same freshly chipped wood has been shown to inhibit the development of mycorrhizae during reforestation. Negative effects on mycorrhizae must be avoided in the landscape because they are very important in the maintenance of healthy plants.

Compost and mulch producers, landscapers, and homeowners can take measures to minimize fungal problems in the landscape. The type of mulch used, fresh versus composted mulch, the moisture content of the mulch before and during its utilization, the temperature and pH of the mulch before and during utilization, and the depth to which it is applied all play a role. Each factor is discussed here.

DISPOSAL OF DEAD AND DYING TREES

Once the trenches are in place, diseased and dead trees should be removed as soon as possible by cutting them down to leave a 2-4 inch high stump. Because diseased trees with bark tightly attached may produce or harbor fungal mats, they should be disposed of promptly. Once the bark becomes loose or sloughs off, no mats can be produced and movement of the infected wood out of the diseased area is no longer a concern in an urban context. (However, strict restrictions apply for movement of diseased wood out of state and internationally.) Thus, either the trees are debarked mechanically, or the timber can be sold to a sawmill for cutting or chipping.

Although no studies are known on the transmissibility of C. fagacearum via wood chips, the pathogen does not survive when exposed to desiccation and is very temperature sensitive. Composting the chips would further reduce or eliminate the pathogen. Thus, it is highly unlikely that wood chips will spread the disease. However, good precautionary practice suggests to avoid using infested chip mulch around healthy oaks.

If the wood cannot be disposed of as described above, it can be cut and split for firewood. Because this process does not involve debarking, firewood can still potentially harbor fungal mats and thus attract Nitidulids during the summer in which the trees died. The wood must be arranged in stacks and covered with 4 mil plastic tarp through the winter (if the wood is used then) or the end of the next season (Oct. 1 of the year following the death of the trees).

By producing a greenhouse effect, tarping will kill the temperature-sensitive pathogen and prevent the beetles from accessing potential fungal mats. Tarping should be done with transparent plastic to produce the desired greenhouse effect. However, black plastic will also work, by concentrating the sun's heat.

In both cases, the best results are achieved by placing the tarped pile in an un-shaded, possibly sunny area. When covering the pile, the tarp should be sealed to the ground to prevent beetles from accessing the pile. For this reason, all punctures in the tarp should be mended with duct tape. At the end of the second season the wood can be safely uncovered and disposed of as preferred, since it no longer constitutes a threat.

Chemical Treatments

Chemical treatments are usually not warranted, due to the high cost of intervention. However, application of systemic fungicides is an option when highly valuable trees are threatened by infected neighbouring trees, or whenever a high risk of infection exists.

High value may be attributed to individual trees or groups of trees in communities, or it may apply to individual trees in a homeowner's yard.

Systemic fungicides have been demonstrated to be effective, particularly when applied as a preventative treatment.

The only scientifically tested systemic fungicide showing any efficacy and labeled for use against oak wilt is propiconazole, available under the trade name Alamo. If a decision is made to apply propiconazole, it must be done strictly according to label and should be carried out by experienced, professional tree service personnel. The product is injected directly into the sapwood on root flares just under the soil line.

This is the best guarantee that the fungicide will be translocated throughout a tree, thus affording the maximum possible protection. In consideration of the disease cycle, the best time of the year to inject trees is early spring. However, application should occur as soon as the risk to a tree is realised, even if it is later in the growing season.

Depending on the tree size and value, treatments should be applied every 12-36 months, with annual assessments. This treatment has virtually no hope of succeeding in infected red oaks, even in early infection stages.

Chemical treatment has a higher chance of success, but only as a palliative measure, with the more tolerant white oaks. In this group, application of the fungicide to trees in early infection stages can result in delay of symptoms and eventual death. It will not, however, rid an infected tree of the pathogen.

MULCH TYPE AND FRESH VERSUS COMPOSTED MULCH

Wood products from some trees are more resistant to decay than others and, therefore, cause fewer problems. Bark chips (nuggets) from large mature pine or other softwood trees such as cypress trees contain mostly lignin (dark material in bark), wax and protected cellulose that resist decay. On the other hand, wood wastes from these same tree species, but ground as young trees, rot quite readily because the cellulose in such bark and wood products is not yet protected from decomposition by lignin waxes or tannins.

Hardwood tree bark (oak, maple, etc.), even from large trees, contains a large concentration of cellulose that is not protected from rotting. Therefore, hardwood bark mulches, like ground wood from almost all tree species, rot readily and cause most of the nuisance mold problems in the landscape. The finer the product is ground, the more severe the problem can be! These materials are low in nitrogen content.

The fine particles (less than 3/4" diameter) in such mulches cause nitrogen immobilization in soil. The microflora that decomposes the wood particles takes up the nitrogen required for growth of plants. The result is that the plant becomes starved for nitrogen. Some mulch producers screen all particles smaller than 3/8" out of high-wood-content or hardwood-bark mulches, which avoids most of the nitrogen immobilization problem. The best way to avoid all these problems and bring about beneficial effects by mulching is to add nitrogen to woody and hardwood bark products followed by composting to lower the carbon to nitrogen ratio. Blending of grass clippings with wood wastes before composting is one way to achieve this. Addition of poultry manure or urea to supply 1.2 lbs. available nitrogen per cubic yard of material satisfies the nitrogen need also. Some landscapers add 10-15 per cent by volume composted sewage sludge to hardwood bark or wood wastes, and this makes an ideal product that has performed very well in landscapes.

These amended products should be composted at least six weeks. This process kills plant pathogens, eggs of insect pests, and produces a nitrified product that releases plant nutrients rather than ties up nitrogen. As mentioned above, the microorganisms that have colonized these products reduce the potential for growth of nuisance fungi and provide control of many plant diseases.

Temperature, Moisture Content, and pH

Landscapers often apply quality mulch products from high temperature piles (140-160 degrees F) directly into the landscape. The temperature of the mulch is high because of heat produced by growth of microorganisms, known as thermophiles, in storage piles during the composting process. These microorganisms die soon after the mulch cools to 50-80oF after it has been placed around homes.

Because they require high temperatures to survive, they cannot grow and compete with soil microorganisms at the low temperature of mulches in the landscape. The sudden temperature drop that often occurs after mulch is applied creates what is known as a "biological vacuum." It also can occur during bagging of products at producers of mulches and particularly during dry seasons.

Mesophiles (low temperature soil microorganisms) rapidly colonize such mulches. If the mulch is dry, or dries out to a moisture content below 34 per cent during the first day after it is applied (mulches are dusty below this moisture content), fungi become the primary colonizers. This sets the stage for problems later, and the problem becomes most severe in mulches that are applied too deep in the landscape.

After prolonged heavy rains, the dry material colonized by fungi eventually becomes wet. Dry products stored in bags may also become moist when water produced as a result of microbial activity accumulates along the inner surface in bags. Bacteria then rapidly colonize the fungal white mass to induce the formation of fruiting structures by the fungi. The nuisance toad stools and other fruiting structures appear a few days later. Mold problems occur also when dry products are bagged or applied to dry soils.

Dry composts removed from high-temperature piles occasionally cause mushroom problems in bags and also in soils at nurseries. These moldy products inhibit plant growth in field soils as well as in potting mixes. They also cause wettability problems if dry conditions persist in the soil for a few weeks to give fungi a chance to become the dominant colonizers. Plants do not grow well in such moldy soils. These problems can be reduced by soaking the high-temperature products with water as they are applied in the landscape or bagged. The high-moisture organic matter then becomes rapidly colonized by bacteria during the first few days.

These bacteria compete with fungi to reduce the potential for the development of major mold problems. This strategy has been successfully applied over the past decade to hardwood as well as softwood composts and mulches. It has controlled nuisance problems caused by many fungi in various parts of the United States and abroad.

The pH or acidity of the mulch is another important factor. Sour mulches that give off acrid odors may range in pH from as low as 2.5 up to 4.8. Highly acidic mulches are toxic to most plants and promote the growth of fungi. Bacteria that inhibit fungal growth cannot colonize mulches when the pH is lower than 5.2. The low pH and fungal problems are avoided if the raw material is nitrified and composted as described earlier.

In summary, water applied at the right time during composting, storage, and mulching can solve most of the fungal nuisance problems. It is best to maintain a water content higher than 40 per cent on a total weight basis. Again this allows bacteria as well as fungi to colonize the organic matter, and it sets up competition for nuisance molds. The moisture content of most organic products actually can be raised above 50 per cent and not present excessive weight problems during transport.

What to Do Once the Problems Occur?

Sometimes very little can be done to control nuisance fungi other than to spade the mulch into the surface soil layer followed by soaking with water. Another option is to remove the mulch, place it in a heap after thorough wetting to allow for self-heating to occur (110-140 degrees F). This will kill nuisance fungi. If fresh dry mulch is placed on top of mulch colonized by nuisance fungi, the problems may occur again the following year or even earlier.

The best control strategy for homeowners and landscapers is to purchase composted products low in wood content. Fresh, finely ground woody products should be avoided for many reasons unless composted first. Coarse fresh woody products are much less likely to cause problems unless applied too deep. It is important to soak all mulches immediately after they have been applied.

Generally, mulches should not be applied to a depth greater than two inches. Mulches and composts applied in this manner provide many types of beneficial effects rather than nuisance problems, or worse, plant diseases. Sour mulches should be avoided altogether.

Mycorrhizas in the Urban Landscape

Mycorrhizas are anatomically intimate associations between fine (feeder) roots of plants and some special soilborne fungi. For this reason they are called symbioses. The plant, being capable of producing its own food through photosynthesis, can potentially survive without the fungus. The latter, however, is often completely dependent on the live plant for sustenance; thus, it is a parasite of the plant. The association usually (but not always) results in a nutritional benefit to both the host plant and the fungus: the plant receives extra soil mineral nutrients, such as phosphorus and nitrogen, and extra water absorbed by the fungus growing out into the soil; the fungus obtains sugars and other vital substances from the plant. This kind of two-way beneficial interaction is called a "mutualistic association."

In nature, the vast majority of plants are affected by this kind of association with fungi, to the point that some scientists claim that "the study of plants without their mycorrhizas is the study of artefacts. The majority of plants, strictly speaking, do not have roots, they have mycorrhizas.". In fact, it is now widely accepted that in the course of evolution, colonization of land by plants occurred thanks to these plant-fungus associations.

Mycorrhizas are important because plants grown in the absence of their fungal associates generally do not perform as well as those that are mycorrhizal. Sometimes, such as in the case of orchids, the seeds do not even germinate in the absence of the appropriate fungus.

Mycorrhizas are also known to protect the plants from various root diseases. However, lack of mycorrhizal roots on normally mycorrhizal plants is a rather rare occurrence, as most natural soils contain a full complement of mycorrhizal fungal propagules, that is, spores or other structures used by the fungi to infect new plants.

MYCORRHIZAS COME IN DIFFERENT TYPES

There are seven different kinds of mycorrhizas. The most important ones from an urban landscape perspective are the ectomycorrhizas, the endomycorrhizas, and the ericoid mycorrhizas. The main and most visible characteristic of ectomycorrhizas is the fungus growing on the surface (hence "ecto," which means "outside") of the feeder roots. This fungal growth is called a sheath or mantle.

The fungus grows from the mantle into the cortical layers of the root, but remains between the plant cells and does not penetrate them, forming a so-called Hartig net. This is where the nutrient exchanges between the plant and the fungus occur. On the other hand, endomycorrhizas are characterized by the absence of a mantle and by penetration of the fungus into the cortical cells.

Once inside the cells, the fungus produces characteristic structures, called "arbuscules" (= tiny trees) and "vesicles." The arbuscules are the structures where the nutrient exchanges between the plant and the fungus occur. Hence, endomycorrhizas are also called "vesicular arbuscular mycorrhizas" (VAM) or just "arbuscular mycorrhizas" (AM). Ericoid mycorrhizas are somewhat similar to VAM, in that the fungus invades the host cells, but affect only plants in the family Ericaceae.

Different plant groups tend to form specific types of associations. For example, Ericaceae tend to form ericoid mycorrhizas, P inaceae almost always form ectomycorrhizas, and hardwoods can form both ecto- and endomycorrhizas, sometimes on the same plant. Table shows a partial list of forest and ornamental trees and their specific type of association. One last characteristic to bear in mind, particularly when considering the following paragraph, is that the fungi involved in the two main types of mycorrhizas, that is endo- and ectomycorrhizas, feature remarkably different levels of host species specificity.

While most individual ectomycorrhizal fungal species tend to form associations with specific host plant species or a restricted number of them, individual endomycorrhizal fungal species are "generalists" and can associate with hundreds of different host plant species. This characteristic is reflected in the number of recognized mycorrhizal fungal species: thousands of ectomycorrhizal vs. about 200 endomycorrhizal fungal species.

Artificial Inoculation

This question is becoming more and more common as more companies have recently begun commercializing mycorrhizal inoculum. Artificial mycorrhizal inoculation may benefit plants only when there is no natural and appropriate mycorrhizal inoculum in the soil, or the inoculum level is low, or species present are less efficient at aiding the plant host than those being introduced. This situation is most commonly found in highly impoverished soils that do not support natural fungal growth, for example in the case of mine spoils.

In principle, highly compacted, organic nutrient-poor soils in urban environments (often the result of mineral subsoil horizons having been turned into the top layers by construction activities) may also be prime candidates for artificial mycorrhizal inoculation. However, success of the treatment can never be guaranteed for two main reasons:

- The urban soil may already contain a complement of mycorrhizal fungi, however depleted, that out-compete the newly introduced species. Competition among mycorrhizal fungi is well documented and a common occurrence in natural soils, for example in reforestation operations. It is particularly important for ectomycorrhizal fungi;
- The soil organic matter content may not be adequate. The presence of organic matter can greatly enhance the development of mycorrhizas.

Due to the specificity properties of the different mycorrhizal associations described above, it is likely that mycorrhizal inoculation will be more successful, in terms of actual establishment of the inoculated species, with endomycorrhizal fungal species, even though the main ectomycorrhizal inoculum being commercialized is a "generalist.

Among specialists." At present, there is very limited, unbiased scientific evidence that artificial mycorrhizal inoculum in the landscape makes plant establishment more successful or that the inoculated plants grow better over time. In fact, the available evidence is rather inconsistent. In part this is due to the fact that commercial preparations may not be very viable or not viable at all by the time they reach the consumer, or become not viable once they are in the consumer's hands.

But perhaps the biggest factor is that soil conditions in the landscape may be too variable to be able to make generalizations about the usefulness of artificial inoculation with the few selected species available commercially. Furthermore, any short-term positive effects observed on plant growth following mycorrhizal inoculation may simply be the result of the fertilizer components that are often present in the commercial preparations. The bottom line is that artificial inoculation may have either a positive or neutral effect on plant growth and health. In all cases, however, it is very unlikely that artificial inoculation is detrimental to the plant.

In other words, the treatment may or may not work, but would not be injurious. Current research, conducted in the author's and other researchers' laboratories, suggests that organic soil amendments, particularly the addition of composted mulches, greatly enhances the mycorrhizal status of landscape plants, even without the addition of artificial inoculum. Improvement of soil conditions may therefore be the most important factor in mycorrhizal development in the urban landscape.

Oak Wilt

Oak wilt is a serious and often deadly vascular disease of oaks. The fungal pathogen, Ceratocystis fagacearum, is believed to be native to the United States and is distributed throughout the Midwest and Texas.

What the Pathogen Does

The fungus grows into and throughout the water conductive tissues (that is, the sapwood) of the host. The fungus plugs the vessels with its own body (mycelium) and spores, but it also causes a defensive reaction by the tree to stop the fungal spread by actively plugging its own vessels. These processes interfere with water uptake and cause a wilting syndrome which often results in death of the tree.

Susceptible Oaks

All oaks are susceptible. Those in the red-black oak group (black, blackjack, pin, northern and southern red, scarlet, shingle and shumard oak) are extremely susceptible and can die within a few weeks of infection. Oaks in the white group (bur, chinquapin, post, swamp white, and white oak) are more tolerant of the disease and may survive infection for one or more years while displaying decline symptoms.

Diagnostic Symptoms

Symptoms are typical of wilts. Leaves usually begin withering in the upper canopy, producing "flags," that is, whole branches or crown portions turning red-brown. Leaves of red oaks typically show yellowing and browning of the leaf margins. White oak leaves usually show rather non-descript symptoms. Conversely, live oaks in the southern United States produce characteristic dead areas along the leaf veins.

These dead areas generally expand until the whole leaf becomes brown. Eventually the leaves fall from the tree. If infections occur in late spring, trees usually begin wilting in mid-summer to late summer, when the plants often are subjected to water deficit due to increased transpiration demand and decreased rainfall.

A specific and sufficient diagnostic character is the appearance on dead and dying red oaks (but not white oaks) of spore-bearing fungal mats under the desiccating bark. These fungal mats crack the bark open with pressure pads to facilitate dissemination of the pathogen. Sapwood streaking is also a good, but insufficient, diagnostic character.

Disease Cycle and Conditions Favouring Disease

In order to properly manage oak wilt it is essential to understand its cycle. The pathogen spreads from diseased to healthy trees in two ways: overland and underground. Overland spread is mediated mainly by sap feeding (a.k.a. picnic) beetles (Coleoptera: Nitidulidae). However, there is some evidence that oak bark beetles (Coleoptera: Scolytidae) may also be involved. Nitidulids are attracted by chemicals emanating from the fungal mats described above.

Once on the mats, the beetles pick up fungal spores and can carry them, sometimes over distances of a few miles, to freshly wounded healthy trees (attracted by the smell of fresh sap). This results in new infections, thus closing the overland cycle. While insect spread is an important medium to long range dispersal mechanism for this fungus, it is estimated that 90 per cent of new infections occur between neighbouring trees through root grafts. In this case, the fungus grows down the trunk, into the roots of diseased trees, and then into healthy trees via the common root system. Once in the new tree the pathogen grows throughout the vascular system and spreads to other trees via the root system or the beetles. In this way, spread through root systems often results in disease centres that expand outward from the initially infected tree.

From the above, it follows that conditions favouring disease include the availability of susceptible oak species, trees growing close to each other, and the availability of fresh wounds for beetle-mediated infection. Pruning wounds are obvious culprits, but any fresh wound will function as potential infection gateway. The word fresh is emphasized because it is believed that wounds are attractive to Nitidulid beetles only for up to three days. As with many plant diseases, other stresses (for example drought) can predispose trees to faster symptom development, and thus worsen the syndrome.

Control and Management of the Disease

The best control for oak wilt is through preventative measures that interrupt the disease cycle.

Prevention of Overland Spread

Overland spread can be hindered or interrupted by ensuring that trees are never wounded between April 15 and July 1. This is when most Nitidulid beetles fly to locate fresh sap and/or fungal mats. A more stringent approach is to avoid wounding the trees throughout the growing season, since additional summer flights of the beetles are possible.

If pruning is absolutely necessary during the growing season, it is imperative to dress the wounds. This can be done with latex paint. Although this will slow wound healing, it will also deter beetles from landing on the wounds.

Prevention of Underground Spread

Given the higher significance of underground spread, control of direct tree-to-tree transmission is much more important. Here, interruption of the disease cycle is accomplished by physically severing actual or potential root contacts between diseased and healthy trees. This is done by trenching or cutting through the soil with a trencher or vibratory plow. The latter is the preferred tool. Given the depth of oak root systems, it is advisable to use a 5 ft blade.

Trenching must always be done before the diseased or dead trees are cut for removal, to avoid sudden water tension imbalances that might "suck" fungal material from the infected trees into the healthy trees through the common root system.

Trenching should be conducted by advice of specialists. This is due to the importance of locating the trenches appropriately between diseased and healthy trees.

When possible, a double trench defining a buffer band of apparently healthy trees between diseased and uninfected trees should be used.

On residential or commercial properties, always determine the location of buried utility lines which may affect the ability to completely sever the graft. Furthermore, walkways, paths, and roads must also be considered appropriately, as tree roots commonly grow under them.

Due to all of these potential obstacles to proper trenching, it is advisable to undertake such operations under the supervision of tree care professionals with expertise in the management of oak wilt.

There is currently no evidence that the blade will spread the pathogen. However, it is good precautionary practice to spray the blade to runoff, between trenches and between plots, with an antiseptic such as Lysol or a 20 per cent bleach solution.

Bibliography

Alfred Steferud: *Diseases of Fruits and Nuts,* Biotech Books, Delhi, 2005.

Bagdi, Madan Lal : *Physiology, Biochemistry and Biotechnology,* Manglam Pub, Delhi, 2007.

Bahar A. Siddiqui and Samiullah Khan: *Plant Breeding Advances and in vitro Culture,* CBS, Delhi, 1997.

Barnett, S A : *Principles of Ethology and Behavioural Physiology,* J V Pub, Delhi, 2006.

Boyer , J.S.: *Measuring the Water Status of Plants and Soils,* Academic Press, N.Y., 1995.

Chadha K. L. and Pareek O. P.: *Advances in Horticulture: Fruit Crops,* New Delhi, Malhotra Publishing House, 1993.

Clark, R B : *Physiology of Crop Production,* International Book, Delhi, 2007.

Collymore L.: *Fruit Production in Barbados,* Port of Spain, Trinidad and Tobago, 1996.

Coste R.: *Coffee: the Plant and the Product,* London, MacMillan, 1992.

Dabholkar, A.R. : *General Plant Breeding,* Concept, Delhi, 2006.

Daubenmire, R. F.: *Plants and Environment,* New York, Wiley, 1947.

David White: *The Physiology and Biochemistry of Prokaryotes,* Oxford University Press, Delhi, 2007.

Dharamvir Hota: *Modern Biotechnology in Plant Breeding,* Gene Tech Books, Delhi, 2007.

Dodia, DA ; IS Patel and GM Patel: *Botanical Pesticides for Pest Management,* Scientific, Delhi, 2008.

Doijode S. D.: *Seed Germination in Fruits,* New Delhi, Malhotra Publishers, 1993.

Fageria, N K ; V C Baligar and R B Clark: *Physiology of Crop Production,* International Book, Delhi, 2007.

Ferentinos L.: *Proceeding of the Sustainable Taro Culture for the Pacific Conference,* Honolulu, HITAHR, 1993.

Ghosh, G K : *Biopesticide and Integrated Pest Management,* APH, Delhi, 2009.

Gowen S.: *Banana and Plantains,* London, Chapman Hall, 1996.

Gupta, O.P. : *Water in Relation to Soils and Plants : With Special Reference to Agriculture,* Agrobios, Delhi, 2002.

Herminie Broedel Kitchen: *Soils and Crops : Diagnostic Techniques,* Satish Serial Publishing, Allahabad, 2004.

Jain, J.B. and Sumit Jain: *Biotech's Dictionary of Plant Breeding and Genetics,* Biotech, Delhi, 2005.

Jain, Sumit : *Biotech's Dictionary of Plant Breeding and Genetics,* Biotech, Delhi, 2005.

Jones, R. M.: *Plant Resources of South-East Asia,* Wageningen, Pudoc Scientific Publishers, 1992.

Kannaiyan, S. : *Rice Management Biotechnology*, Associated, Delhi, 1995.

Kapoor, R.L. and M.L. Saini: *Plant Breeding and Crop Improvement*, CBS, Delhi, 1997.

Kataria, T N : *Plant and Crop Physiology*, Pearl Books, Delhi, 2008.

Khan, S.: *Plant Breeding Advances and in vitro Culture*, CBS, Delhi, 1997.

Lal. Madan : *Physiology, Biochemistry and Biotechnology*, Manglam Pub, Delhi, 2007.

Mishra, Shubhrata R. : *Morphology of Fungi*, Discovery Pub, Delhi, 2005.

Mishra, U K : *Physiology of Animal Growth and Bioenergetics*, Satish Serial Pub House, Delhi, 2008.

Mitra S.: *Postharvest Physiology and Storage of Tropical and Subtropical Fruits*, Oxon, CABI, 1997.

Nobel, P. S.: *Physicochemical and Environmental Plant Physiology*, Academic Press, San Diego, 1999.

Oldham P.: *Cost of Production of Major Tree Crops in Dominica*, Roseau, Ministry of Agriculture, 1991.

Pareek O. P.: *Advances in Horticulture: Fruit Crops*, New Delhi, Malhotra Publishing House, 1993.

Parry M.L.: *Climatic Change, Agriculture and Settlements*, Dawson Folkestone UK, 1978.

Politycka, A.R. and C L Goswami: *Plant Physiology : Research Methods*, Scientific, Delhi, 2007.

Politycka, B and C L Goswami: *Plant Physiology : Research Methods*, Scientific, Delhi, 2007.

Ramann, E. : *The Evolution and Classification of Soils*, Asiatic Pub, Calcutta, 2006.

Rao, P. Venkateshwara : *Dairy Farm Business Management*, Biotech Books, Delhi, 2008.

Saini, M.L. : *Plant Breeding and Crop Improvement*, CBS, Delhi, 1997.

Sathe, T V : *Biotechnological Approaches in Entomology*, Manglam Pub, Delhi, 2008.

Sharma, R.C. : *Diseases of Horticultural Crops: Fruits*, Indus, Delhi, 1999.

Sharma, Rajni : *An Introduction to Plant Morphology*, Campus, Delhi, 2004.

Shivanand Tolanur: *Practical Soil Science and Agricultural Chemistry*, International Book Distributing, Delhi, 2004.

Singh, S K : *Biotechnology, Plant Propagation and Plant Breeding*, Campus Books, Delhi, 2008.

Tripathi, Pramila : *Botanical Pesticides in the Management of Post-harvest Fruit Diseases*, Daya, Delhi, 2005.

Trivedi, P C : *Plant Physiology : Current Trends*, Pointer, Delhi, 2007.

Tyagi, I.D. : *Plant Breeding and Genetics at a Glance*, South Asian, Delhi, 2005.

Vanderplank, J.E.: *Plant Diseases: Epidemics and Control*, New York, NY, USA, Academic Press, 1963.

Verma, L.R. and R.C. Sharma: *Diseases of Horticultural Crops: Fruits*, Indus, Delhi, 1999.

Walia, S. and B.S. Parmar: *Pesticides, Crop Protection and Environment*, Oxford & IBH, Delhi, 1995.

Whealy K.: *The Garden Seed Inventory*, Decorah, Seed Saver Publications, 1988.

Index

R

S

V

W

❑❑❑